Stochastic Models for Fault Tolerance

Katinka Wolter

Stochastic Models for Fault Tolerance

Restart, Rejuvenation and Checkpointing

 Springer

Katinka Wolter
Institute of Computer Science
Freie Universität Berlin
Takustr. 9
D-14195 Berlin
Germany
katinka.wolter@fu-berlin.de

ISBN 978-3-642-43500-3   ISBN 978-3-642-11257-7 (eBook)
DOI 10.1007/978-3-642-11257-7
Springer Heidelberg Dordrecht London New York

ACM Computing Classification (1998): C.4., G.3, D.4.8

© Springer-Verlag Berlin Heidelberg 2010
Softcover re-print of the Hardcover 1st edition 2010

Cover design: KuenkelLopka GmbH, Heidelberg

Printed on acid-free paper.

Springer is part of Springer Science+Business Media (www.springer.com)

To Tobias and our children, Louise, Margarete, Elisabeth, Samuel and Jakob

Foreword

The book you hold in your hands, entitled Stochastic Models for Fault Tolerance - Restart, Rejuvenation and Checkpointing, found its incarnation in the hidden wonders of the Internet. On the face of it, the Internet may look like just any other network, such as the telephony network of old, cable TV network, satellite communication networks or bank transfer networks. However, when considering the actual usage patterns of the Internet, an exciting world of mathematical interest and curiosity opens up, to which that of other networks pale in comparison. Caused by the sheer number of users and web services, as well as the elegant intricacy of the packet-based network technology, measurements of the Internet have revealed highly intriguing patterns. These patterns exhibit such phenomena as the small world effect, scale free networks and self-similarity, each of which one can find discussed extensively in popular and scientific literature alike.

The work reported in this book came about because of another interesting Internet phenomenon, namely that of the 'heavy tail'. It says that high download times are relatively common, much more common than with the thin exponential tail that characterizes completion times in traditional communication networks. This has a fascinating consequence, which under certain assumptions can be proven mathematically: it is often faster to abort and retry a download attempt than to wait for it to complete. After all, one may have been caught in the heavy tail, from which one can only escape by clicking the reload button. This fascinating fact seduced us into conducting research into the optimal timing of these retries (or restarts as they will be called in this book), eventually leading to Katinka Wolter's Habilitation and this book.

Our research resulted in a number of interesting theoretical results, accompanied by extensive experimental work (mostly carried out by Philipp Reinecke). In this book, Katinka Wolter embeds these results into an extensive survey of existing scientific literature in restarts, rejuvenation, checkpointing and preventive maintenance in general. These categories of techniques have in common the problem of timing: how frequent should one carry out the preventive or pre-emptive activity? Mathematically, this leads to a set of related problems and solutions, and this book provides the reader with a careful overview of the various mathematical results as well as their inter-relations.

Many authors have researched problems of timing with great success, and the book highlights the various contributions. From a historical perspective, computing related research on timing was preceded by that on preventive maintenance of systems in general, such as for engine parts, machinery, and so forth. The results for computing systems build on these generic insights, roughly in three phases. First, mostly in the eighties of the past century, the question of when to save intermediate results (a 'checkpoint') became important. If a long-running computation fails, a checkpoint assures one does not need to start from scratch, but can roll back to the last checkpoint. The mathematical problem then is to compute the checkpoint interval that minimizes the completion time of the computation, under various fault and load assumptions.

In the nineties, researchers became interested in rejuvenation, especially when it was proposed as a generic fault tolerance approach by Bell Labs researchers. In rejuvenation, a computer will be switched off and on deliberately to avoid imminent more serious uncontrolled crashes. The question is when to 'rejuvenate' the computer. This can be based solely on a timer, but as the survey in this book shows, rejuvenation models tend to become more complex since the key for this problem is to identify the state in which to rejuvenate the system. Finally, retry problems, such as in Katinka Wolter's main line of research, have been discussed throughout the history of computing. In some sense, retries are the dual of preventive maintenance: both deal with the life time of a process, but maintenance tries to prolong life (of a system), while retries try to shorten life (of a job) as fast as possible. This fundamental difference introduces subtle differences in the mathematical models and makes that problems of interest in preventive maintenance do not always translate to retries. In the respective chapters, this book will illustrate these differences.

Readers of this book will find an up-to-date overview of the key theoretical results for timing restarts, rejuvenation and checkpointing, thus serving as a hand book for engineers facing this issue. Computer researchers may find inspiration in the elegance of many of the results, and contribute their own extensions or improvements. Non-specialist readers may hopefully catch a glimpse of the wonders hidden inside the Internet, which emerge from the combined behaviour of all of us browsing the web, downloading music and emailing friends. For all readers, the next time you are browsing the web and a page takes long to download, you may think of this book and use it to time your clicking of the reload button.

It has been a sincere pleasure to have been associated with the research that eventually accumulated into this book. It was great to work with Katinka as well as Philipp solving some of the research problems. I congratulate Katinka Wolter with completing the challenging but extremely valuable task of writing this book and trust the result will be solid information and occasional enjoyment for many readers.

Newcastle, UK,   Aad van Moorsel
October 2009

Preface

As modern society relies on the fault-free operation of complex computing systems, system fault-tolerance has become a matter of course. Common agreement exists that large software systems always contain faults and precautions must be taken to avoid system failure. Failure of hardware components often is caused by external factors that can be neither predicted, avoided, nor corrected. Therefore, mechanisms are needed that guarantee correct service in the presence of failure of system components, be it software or hardware elements. Commonly used are redundancy patterns. These can be either redundancy in space or redundancy in time. Redundancy in space means the inclusion of additional hardware or software modules in the system that can replace a failed component. Different types of redundancy in space exist, such as cold, warm and hot stand-by. Redundancy in time includes methods such as restart, rejuvenation and checkpointing, where execution of tasks is repeated, or the task environment is restarted, either following a system failure or, preventively, during normal operation.

This book is concerned with methods of redundancy in time that need to be issued at the right moment. In particular we address the question of choosing the right time for the different fault-tolerance mechanisms. This includes a brief introduction to the regarded methods, i.e., restart, rejuvenation and checkpointing and aspects of their practical implementation in real-world systems. But the focus of the book is on selecting the right time, or timeout, for restart, rejuvenation and checkpointing. In general, this is the timeout selection problem.

Selecting the right timeout is a problem that is subject to a number of uncertainties. It is, in general, unknown when the system will fail. Furthermore, upcoming busy and idle periods of the system are not known, as is future user interaction with the system. Due to the many uncertainties, the timeout selection problem lends itself for a stochastic treatment. Consequently, many stochastic models addressing the timeout selection problem in restart, rejuvenation and checkpointing have been proposed. This book gives an overview of existing stochastic models of restart, rejuvenation and checkpointing.

The second part of this book treats a stochastic model of restart in various facets. Restart operates on program, or application level. If a task does not complete within a given time it is suspected to have failed and is, consequently, aborted and restarted. The timeout after which to abort and restart the task must be carefully chosen because if it is too short the task might be aborted just before completion while if it is chosen too long one must wait unnecessarily.

The second method is software rejuvenation. Software rejuvenation restarts the operating environment of a task in order to prevent failures. Rejuvenation is a proactive fault-tolerance treatment. Hence it is issued before the system fails. This is particularly challenging, as it implies assumptions as to when the system would fail if no measures were taken. Ideally, the rejuvenation interval would always end just before the system fails. A conservative choice of the rejuvenation interval will select short intervals. But rejuvenation comes with a cost of saving the operating environment and all processes, restarting the system and reinitialising the operating environment and all processes. If rejuvenation is performed too often the rejuvenation cost accumulates unnecessarily, while if the system is rejuvenated at too long intervals it will often fail, rendering rejuvenation uneffective.

The third method, checkpointing, is the most complex mechanism of the three as it has a preventive component, saving a checkpoint, and a reactive component, rollback recovery. Checkpointing systems save the system state in regular or irregular time intervals. Upon failure the system recovers by rolling back to the most recent checkpoint. The work performed since the most recent checkpoint is lost with a failure. If checkpoints are taken too frequently the interrupt and save operation incurs too high a cost, while if checkpoint intervals are too long much work is lost upon system failure.

For all three methods similar trade-offs exist. The fault-tolerance mechanisms come at a cost that must be traded against the cost of a potential system failure. If the timeouts are well chosen the fault-tolerance mechanism will avoid a failure and be worth-while. The trade-offs can be evaluated and optimised using stochastic models. The focus of this book is to collect, summarise and compare those stochastic models. This can be seen as a first step towards understanding and solving the generic timeout selection problem.

This book is based on the author's habilitation thesis at Humboldt-University in 2008. The habilitation thesis, and hence this book, would not be as it is without the careful and thorough reading from the first to the last page of Mirek Malek. I would like to thank him for his efforts. His many valuable comments helped to improve this text tremendously. I am thankful to Boudewijn Haverkort and Miklos Telek as they agreed on reviewing and commenting on the thesis. Miklos Telek even came to Berlin for the habilitation lecture, even though it was on yet another topic.

I thank the members of the habilitation committee for their invested time and consideration. It was a pleasure to work with the gifted students at Humboldt-University over the five years I spent there until completion of my habilitation thesis. I particularly thank Willi Engel for being a member of the habilitation committee and Philipp Reinecke for the many little shell scripts he quickly set up and for his dedicated work in the restart project. I thank Steffen Tschirpke for technical support and Christine Henze for help in all administrative matters. I am grateful to my former colleagues at Humboldt-University for many fervid discussions that broadened my view.

Aad van Moorsel has been a colleague on the restart project for many years. He is the perfect person to write a foreword and I am happy he agreed to do so.

I want to take the opportunity to thank Jochen Schiller and his group for giving me such a warm welcome and making me feel at home immediately at Freie Universität. This made me anticipate future work related to restart and also broadened my horizon.

I would like to thank Springer Verlag and Ralf Gerstner in particular for agreeing to publish this book.

Finally, the most crucial support throughout the past years came from my family. I thank Tobias Zepter and our children Louise, Margarete, Elisabeth, Samuel and Jakob just for being there and being my delight every day.

Berlin, Germany,   Katinka Wolter
October 2009

Contents

Part I
Introduction

Chapter 1
Basic Concepts and Problems

This book adresses problems and questions in computer fault-tolerance that can be tackled using stochastic models. Computer fault-tolerance is an important feature in mission-critical or highly available systems. It is implemented through a system design that includes replication and redundancy [127]. We are interested in problems related to redundancy rather than replication, namely the question when to issue a certain methods. Redundancy mechanisms can be divided into redundancy in space, where a stand-by component may take over service, and redundancy in time, where the service is repeated. We will more closely consider the latter and study different methods that have in common that they all implement some mechanism of repeating a task. Recent studies [68, 117] show that modern software systems are far from being fault free and the faults causing a system failure often can be localised. Even more, the Pareto distribution of faults [58] holds, i.e. 80% of the faults are located in 20% of the files. Frequently, failure of the system is caused by a number of small faults that cannot be removed. In this situation restarting modules may be the method of choice.

Even though there are many interesting open issues in software fault-tolerance the focus of this book lies on the timeout selection problem rather than on the many other aspects of fault-tolerance mechanisms using redundancy in time. Before diving into a treatment of stochastic models adressing the timeout selection problem for the three selected methods, i.e. restart, rejuvenation and checkpointing, this chapter discusses preliminary aspects of the general timeout problem common to all further algorithms.

1.1 The Timeout Problem

Timeout problems arise virtually in all walks of life, including human-created systems such as computing systems. Every day often several times we find ourselves waiting for some service, or waiting for a delivery, wondering *how long should I wait?* In computing systems one has to wait for the result of a computation, the response to a request, or a reply in a communication. Set aside the situations where a system deliberately waits for customers. Behind the question, how long one should

K. Wolter, *Stochastic Models for Fault Tolerance*,
DOI 10.1007/978-3-642-11257-7_1, © Springer-Verlag Berlin Heidelberg 2010

wait, lies the suspicion the system may have failed, the length of a queue might have become too long such that the request may have got lost, the connection may have been disrupted or a contributing process may have erroneously stopped and will never reply. In real-life situations the length of the line in front as well as the performance of the person behind the counter can be observed. Based on both impressions the decision whether to stay in line or to drop out is taken. The human user thus applies an educated guess in order to decide whether the waiting time is still within reasonable limits of the system response time in healthy state. This guess can often be more precise in a human environment than when facing a computing system. The less is known about the internal configuration of the system, the more a user has to rely upon his feelings. Typically first the user will wait optimistically for some time followed by a period of doubt, until finally one gives up hope and cancels the session, aborts the computation, or interrupts the communication.

The question how long to wait touches on different fields of computer science. It implies consideration of the reasons for long waiting time and investigation of possible solutions of the problems leading to long waiting times. Waiting time can be long because of a failure somewhere in the system where no appropriate fault-tolerance mechanisms are deployed. One might have to wait because a computation takes much longer than anticipated, or some part of the system is just very slow, or overloaded. Analysis of delays in computing systems requires detailed consideration of system performance and reliability parameters and possible performance and reliability enhancements.

We study different performance and reliability mechanisms commonly used in computing systems. One can do this at different levels of abstraction. Implementation of the mechanisms at system level raises many questions, such as when to use a method, how to implement it, in which layer of the software stack to use what mechanism, how to evaluate each method, how to improve them and how to parameterise them. The last two of those exceed pure systems questions and are commonly answered using formal methods, or mathematical models. This work is concerned with stochastic models and stochastic modelling solutions to some particular timeout problems as they arise in performance and reliability mechanisms.

In the second part, the first technical part of this work, a black box approach is used to find ways to speed up processes. The restart method does not explicitly consider the shortcomings in a system that may lead to long waiting times. Instead, it uses an engineering perspective that does not require an understanding of the system dynamics leading to exceptionally long delays. It uses the system's external timing behaviour to, first, identify situations in which matters can be sped up by aborting a process and starting it anew, and, second, develop algorithms that determine the optimal timeout efficiently. This part consists almost exclusively in recent research contributions of the author and her colleagues.

The method studied in the third part focusses on the impact of the environment on process behaviour. It analyses not the performance of the process itself, but its response to the degradation of the environment. The process environment is analysed, modelled and tuned to improve the process performance and reliability.

Rejuvenation denotes a periodic restart of the process environment in order to increase system performance and reduce the failure probability.

The fourth part is concerned with checkpointing, which is the most fine-grained performance and fault-tolerance mechanism out of the considered approaches. Checkpointed systems periodically save the system state such that in the case of system failure no complete restart of the running processes is necessary, but all processes can be reset to the most recent checkpoint and continue processing from there. The checkpoint intervals can be placed in different ways depending on the considered system and the metrics of interest.

At first sight the mechanisms discussed in this work, restart, rejuvenation and checkpointing, seem to be very similar. In all approaches processing is interrupted and started again. All three methods use a time interval, either as a timeout for process or system restart or for the placement of checkpoints. But even though all methods relate to the general timeout problem each one of them does so in a different way. When using the restart method completion of a process is anticipated as soon as possible and action is taken if the result is not delivered within a given time, whereas rejuvenation and checkpointing are applied to extend the time until system failure. Rejuvenation and checkpointing also speed up task completion but they do so indirectly by avoiding system outage. When applying the restart method not the environment is observed but the considered task. Rejuvenation, in contrast, requires monitoring the system, not the processes it executes and is applied to the system not to the processes. Checkpointing, finally, combines properties of both, restart and rejuvenation. To apply checkpointing the system failure behaviour and the task processing must be observed. The rollback to the most recent checkpoint can be applied to either the task, the system or both. As with rejuvenation the purpose is to circumvent system failures and achieve completion of processes in as short a time as possible. However, the most important difference between restart and both rejuvenation and checkpointing is that the former relates to a minimisation problem with known limit of the optimum, while both latter methods relate to unbounded maximisation problems.

Just as timeout problems in systems at first glance seem to be all the same, the stochastic models appear to be very similar. Only a deeper analysis of the stochastic models formulated for restart, rejuvenation and checkpointing reveals the great impact of the small differences, as will become clearer throughout this work. Analytical models for retrial queues [42, 43, 6] can be seen as more general predecessors the models considered in this text.

A profound understanding of timeout problems in general may provide us with a tool box of appropriate solutions which can be applied after checking some characteristics of the given system. This book takes one first step towards the necessary appraisal of timeout problems by summarising and comparing different timeout mechanisms.

Many threads of the work in this book may be further expanded as they require a separate tractive. Especially development of systems that implement restart, rejuvenation and checkpointing, but also evaluation of such system would be of utmost interest. Bearing in mind that the majority of the modelling work for checkpointing has been published at least 20 years ago and real systems such as IBM's super computer Blue Gene/L and Blue Gene/P implement checkpointing schemes there obviously exists a considerable quantity of research in system development that could supplement the modelling work collected in this book. Recently hardly any work on model-based analysis of checkpointing is being published and the research questions seem to be mostly solved.

For software rejuvenation matters are slightly different. Existing modelling work is roughly 10 years old and until now the method is mainly applied for research purposes. Recently, rejuvenation has been proposed as the method of choice for self-healing systems. Omitting the system's perspective on rejuvenation is less a loss than it is for checkpointing since for the former less work has been published as of yet. Furthermore, the modelling work for rejuvenation often is interwoven with system development and evaluation.

Confronting modelling and model analysis of the restart method with its system implementation leaves yet a different mark. Modelling work on the restart method is quite new as restart has been considered a special case of checkpointing until recently. No dedicated implementations exist. It will be shown that the stochastic restart model is inherently different from known checkpointing models as is their analysis and its results. Even though no systems notably implement the restart method the mechanism is used in many technical systems, the TCP (transport control protocol) being but one example.

Even though the classical timeout models seem to be well understood, the rise of today's adaptive, self-diagnosing, self-healing, and self-managing systems puts the issue of appropriate timeout selection again into question. Hardly any system knowledge is available for heterogeneous, distributed systems and simple, efficient models like the restart model are, therefore, potential candidates for reliability enhancement of such systems. Even though checkpointing models are superior in the amount of system detail they use the restart model is a much more likely candidate in environments where little information is available, but online algorithms are sought that can operate without human interaction. How to set a timeout correctly and efficiently in complex heterogeneous systems is a relevant question already and will become even more so in the future. To be able to answer this question, solid knowledge and profound overview of existing work is essential. To provide a structured collection of existing work in timeout related research is a major contribution of this book.

1.2 System and Fault Models

Restart, rejuvenation and checkpointing apply to computing systems in general. All three methods can be used in processing systems that execute either large tasks which run for a long time, or small transactions that complete quickly. The considered computing systems can be unreliable and subject to failures. In rejuvenation and checkpointing models failure characteristics are of importance while the restart model does not include a model of the computing system explicitly. In a more detailed description each method demands for its own system and fault model. Detailed studies of fault and failure data such as [68] are needed to validate the fault and failure models.

When the unreliable system eventually fails it is being repaired and then restarts operation. As it starts processing again different cases must be distinguished. The failure can be non-preemptive, in modelling terminology, and, in system terminology, the system state has been saved at the time of the failure. The failure can, on the other hand, be preemptive and all work performed so far is lost. Several ways exist to handle the loss of performed work in a stochastic model. The different failure modes have first been studied in unreliable queueing systems [174] and in priority queues [75]. When a *preemptive resume* failure happens the amount of work already performed is memorised. As the system resumes operation only the remaining work of the current task has to be worked off. A *preemptive repeat* failure indicates that all work performed so far is lost as the system fails. Upon resume after a *preemptive repeat identical* failure the task has to be processed again from scratch and the system executes the same task again with identical work requirement. A *preemptive repeat different* failure is followed by reprocessing using a new set of random parameters.

Let us introduce the system and fault models of the three methods studied in this book: restart, rejuvenation and checkpointing.

The restart method simply consists of the abortion of a task and its restart[1] from the beginning. Each execution of the task is independent of all preceding executions, which means that no system state is saved during operation and the used task parameters are initialised independently of all previous initialisations. The restart method uses a very general concept which is being used in reliable communication protocols, it has been applied to portfolio theory in economics [104] and can be applied for virtually any kind of computer science application. Restart has no prerequisites on the system level, other than that the task has not completed yet. Even though the system model for restart does not include an explicit notion of failures the method is applied successfully only if the system experiences transient failures [138]. The type of failure, however, need not be further specified. Transient failures imply that the system may either have recovered when the task is being reexecuted or the failure may not occur again upon reexecution of the task. Note that restart is a method of preventive maintenance and is implemented as timeout mechanism in distributed systems.

Rejuvenation denotes the periodic restart of the operating environment of a process for the purpose of garbage collection, memory clearance, etc. After rejuvenation the system is as new. This is useful under the assumption that over time, e.g., memory leakage degrades system performance and will eventually lead to system

---

[1] The restart method should not be confused with the equally named fast simulation technique [112, 169]

failure. The failures handled using rejuvenation are crash and hang failures due to minor computational errors.

Checkpointing requires saving the system state at intermediate points in time. Upon failure the system rolls back to the most recent checkpoint and computation is resumed from there. If one long batch job is being processed the state of the system including the status of the task are saved in a checkpoint. If checkpointing is applied to a transaction processing system where many short transactions are being processed a checkpoint typically contains only the system state while the status of the transactions in the system is kept separately in the audit trail. Checkpoint and audit trail together allow restoration of the system at the most recent checkpoint and reprocessing of the transactions if necessary. Checkpointing can be applied to long batch jobs, called program level checkpointing, as well as to transaction processing systems, called system level checkpointing[140]. Both give rise to different stochastic models. Checkpointing protects against many transient hardware and component failures. If the fault is caused by a design error the system will fail again with each restoration. Protection from software faults typically requires additional mechanisms such as algorithmic diversity and recovery feed-back. Any kind of permanent failure would disrupt the system at the same point again upon recovery.

The terms retry, restart and reboot each have a specific meaning in fault tolerance theory. In the strict sense retry denotes the identical reprocessing of a task which could be reprocessing a transaction by a running system. Restart means that the task is aborted, the system is shut down and restarted such that a new instance of the system is ready for processing. Reboot includes a system update or reinstallation of the system environment and the operating environment may not be the same as before the reboot. Retry, restart and reboot can all be applied at different levels in a computing system. They intervene at different degree with system operation. While a retry not necessarily requires any process to terminate, restart happens after abortion or termination of a process and reboot require a complete system or component setup.

The abstract relation of retry, restart and reboot is shown in Fig. 1.1. Application restart requires all requests to be retried. Therefore retry is necessary if restart is performed and the set of all retries is a subset of the set of all restarts. Both are

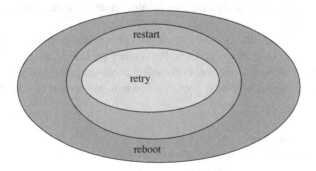

Fig. 1.1 Relation of retry, restart, and reboot

subsets of the set of reboots. In other words system reboot requires application restart and request retry, while application restart only requires request retry and retry is possible without any of restart and reboot.

As a matter of fact, in most cases the restart method in practice retries the current task and rejuvenation applies restart. However, the stochastic models discussed in this book do not deploy the difference between retry, restart and reboot. Mostly only task repetition is modelled and the above discussed failure modes are distinguished abstracting from the circumstances of task reprocessing. Since the distinction between retry, restart, and reboot is for the modelling exercise of minor importance we follow common practice and use terms like restart, reload, retry, etc. interchangingly.

The fault models for restart, rejuvenation and checkpointing are typically very simple. Faults are assumed to be transient and failures are detected immediately as they occur. They are assumed not to have spread in the system. These assumptions are in practice often unrealistic, but they are essential in order to obtain tractable models.

Looking from a different angle, the models studied in this book can be characterised by the metrics that can be computed from them. Of major importance is the task completion time and its moments. Some models allow to draw conclusions on system availability and the cost due to failures or fault-tolerance mechanisms. Very often the cost of a fault-tolerance mechanism is traded against the cost of failures and the purpose of modelling is to minimise the total cost. For all three mechanisms the intervals between restart, rejuvenation, or checkpoints can be chosen such as to optimise the considered metric.

Many different modelling formalisms have been applied for modelling restart, rejuvenation and checkpointing and some of them are presented in this book. For the restart model closed form stochastic expressions are being used while the rejuvenation models often use Petri net models and stochastic processes. An overview of the model classes and their respective solution methods is given in [13]. Checkpointing is modelled using Markov chains, queueing models and analytical closed-form models.

1.3 Preventive Maintenance

Preventive maintenance is a very broad discipline, including observation of the system state and actions that must be taken to prevent failure of the system. Preventive maintenance typically includes observation of the system state through measurements, testing and systematic inspection as well as treatments to improve the actual condition of the system such as partial replacement, component restart, etc. [64, 26]. Knowing what to test and measure and when to take action in order to optimise system operation and avoid failures are the challenges of preventive maintenance. In this book software rejuvenation, one flavour of preventive maintenance, is investigated as it uses a restart mechanism and hence falls into the class of timeout problems. Software rejuvenation requires the monitoring of variables that are indicative of the system degradation such that, if necessary, software restart can be applied as curative treatment. Software rejuvenation stops the system during fault-free operation and performs garbage collection, memory clearance and restart of some underlying processes to avoid potential future failures. Stochastic models for software rejuvenation therefore include a model of the system degradation process as well as model assumptions on the effects of software restart and allow to determine e.g. the optimal time interval between software restarts.

Under the assumption that system degradation follows certain patterns system failures can be predicted and often prevented by means of preventive maintenance in due time [2]. Software rejuvenation as one means of preventive maintenance on one hand causes system downtime, on the other hand it presumably increases system lifetime. The optimal time between preventive maintenance actions can be determined using a stochastic model of the considered system, its degradation and failure patterns. In most systems no well-defined relation between system degradation (e.g. memory usage) and time to failure is known. Then thresholds for presumably indicative system variables must be defined and when these given limits are reached preventive maintenance action is taken. Statistical models can be used to find solutions to the related optimisation problems. A number of different models have been formulated assuming certain degradation characteristics or failure patterns and software rejuvenation properties. Modelling periodic inspections allows to determine an optimal alert threshold, while modelling a degradation process allows to determine the optimal rejuvenation intervals. Other stochastic models are able to optimise the trade-off between downtime due to preventive maintenance and downtime due to system failure.

When searching for common properties of restart, rejuvenation and checkpointing the relationship between restart and rejuvenation is of special interest since restart, like rejuvenation, can be considered a preventive maintenance approach. In particular, extending the system life time can be seen as the dual problem of the completion time problem studied in the next part of this book. In other words, maximise the time to failure through preventive maintenance policies, instead of minimise the completion time through restart policies. Resulting schemes that optimise the timing of preventive maintenance are known as age replacement policies, and the policies discussed for the restart model in this book as well as in [99] are in fact age replacement policies. Interesting enough, it is not easy to find results on this dual model (we have only found one in [57]), bounding the first moment of time to failure, see Sect. 4.3.1.1. In general, preventive maintenance, and rejuvenation in particular, is mostly analysed in terms of cost of preventive versus required maintenance, thus complicating the model, but this is necessary to overcome trivial optimal preventive maintenance solutions.

1.4 Note on Terminology

This book collects, summarises and compares work on stochastic models for restart, rejuvenation and checkpointing. The three methods exhibit interesting relationships, as do the related stochastic models. However, the stochastic models have been applied to different fields in computer science and new application areas may still arise. While the models are all about states and transitions between those states, or activities that direct the model into a new state, the terminology in each of the application areas is inherently different. While from a systems perspective a transaction and a task are certainly not the same thing, in a stochastic model they only differ in their duration. We will therefore often use task, job, work, and transaction alternatingly. Sometimes we may speak of the work requirement, task length or processing requirement and these always refer to the same item in the analysis, the time needed to finish a job. As mentioned above also restart, retry, reboot, and reload have a more general meaning in our context and are therefore not distinguished.

Assuming exponentially distributed time to failure and repair time throughout this book failure of a system happens at rate γ and the repair rate is ν. Commonly in dependability theory λ and μ are being used. In this work transactions arrive to the system at rate λ if they arrive in a Poisson process. They are processed at rate μ for exponentially distributed service time. Other service time distributions are usually specified by their first moment $E[S]$, where S is the random variable denoting the service time. As known from queueing theory $\rho = \lambda/\mu$ or $\rho = \lambda E[S]$. The recovery rate after preventive maintenance as well as checkpoint rollback recovery is denoted r, or R.

The deterministic bound on the downtime, or the accumulated downtime is denoted b, while the random variable denoting the bound on the downtime, or accumulated downtime is denoted T.

If it is finite, the number of allowed restarts is denoted K, as is the number of allowed repairs.

1.5 Outline

This section gives a brief overview of the organisation of this book. This text is structured in four parts.

The first part contains two chapters, the introduction and a chapter on task completion time in unreliable systems. In the first chapter the timeout selection problem is introduced. This is the core problem addressed in this book. Whether restart, rejuvenation, or checkpointing is investigated the provided solution offers an optimal timeout value with respect to some optimisation criteria after which restart, rejuvenation and checkpointing is triggered. The system and fault models underlying the work presented in this book are depicted. Preventive maintenance as a means to enhance software fault-tolerance is discussed. Software rejuvenation, generally, is considered a method of preventive maintenance but also checkpointing belongs to the class of preventive methods. A note on terminology and this outline conclude the first chapter. The second chapter presents models of completion time in unreliable systems without restart, rejuvenation and checkpointing. This is the state upon which the three fault-tolerance mechanisms, restart, rejuvenation and checkpointing ought to improve.

The three subsequent parts are one for each of the considered methods, i.e. Part II covers the restart model, Part III software rejuvenation and Part IV checkpointing.

The restart model in the second part is the authors main research contribution also published in [161, 168, 177, 167, 129]. Work on the restart method is introduced by a brief review of applications of the restart mechanism. Simple restart after a timeout is applied in diverse fields of computing science. This leads to the general question when restart is beneficial to a process and should be applied. This question can be formulate in a straight forward way, but is not easily answered in general. We will give some selected detailed answers in Chaps. 4 and 5 and provide a guideline and simple rule.

Chapter 4 contains our analysis of moments of completion time under restart. This chapter deeply investigates the topic. We look at different moments of completion time and the information the moments provide us with as well as algorithms to optimise the moments when using a finite or infinite number of restarts.

In Chap. 5 we investigate a problem that commonly arises in real-time systems. The question is how to improve the probability of meeting a deadline. How should the restart interval be chosen as to optimise this metric. Furthermore, we develop an engineering rule and algorithm which can be used online.

Software rejuvenation and preventive maintenance in general is discussed in Part III of the book. First, pragmatic aspects of systems implementing software rejuvenation or other methods of preventive maintenance are addressed in Chap. 6. Chapter 7 revisits and compares stochastic models of preventive maintenance and software rejuvenation.

In Part IV work on checkpointing is presented. As in Part III the first chapter in this part, Chap. 8, gives an overview of systems implementing the method. In Chap. 9 stochastic models for checkpointing at program level and then stochastic models for checkpointing at system level are presented. Attention is drawn to some potential improvements of models taken from the literature.

Chapter 10 summarises the material, concludes this book and points out possible directions of future research.

Chapter 2
Task Completion Time

As a reference model for the reliability enhancement techniques discussed later in this book we will first look at system performance, availability and reliability without restart, rejuvenation, or checkpointing. Task completion time is considered in general [15] in unreliable systems that are subject to failures. The three mechanisms can be evaluated by how much they improve the task completion time at what cost. To be able to value the benefit obtained by reliability enhancement mechanisms knowledge of the system behaviour without those mechanisms necessary.

Figure 2.1 shows the generic system model used throughout this book, if not stated otherwise. The system alternates between an up (or operational) and a down (or non-operational) state according to a semi-Markov process. Both the uptime (U), or time to failure, and the downtime (D), or time to repair are random variables with probability distribution function (PDF) $F_U(t)$ and $F_D(t)$ and probability density functions (pdf) $f_U(t)$ and $f_D(t)$, respectively. Transition from the up to the down state and back to the up state occur according to the respective hazard rate functions $h_U(t)$ and $h_D(t)$. This model includes the special case of a Markovian fault model, where failures occur in a Poisson process and repair times are exponentially distributed. Then both, $F_U(t)$ and $F_D(t)$ are exponential distributions and the hazard rates evaluate to the rates of the respective distributions, i.e. $h_U(t) = \gamma$ and $h_D(t) = \nu$. The chosen notation attempts to avoid confusion with the arrival and service rate in processing systems, λ and μ, discussed later.

Successive sojourn times in the up or the down state are always assumed iid. In general, the distribution function and the hazard rate function relate as

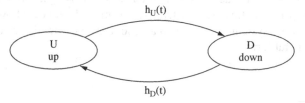

Fig. 2.1 System model

K. Wolter, *Stochastic Models for Fault Tolerance*,
DOI 10.1007/978-3-642-11257-7_2, © Springer-Verlag Berlin Heidelberg 2010

$$F_X(t) = 1 - e^{-\int_0^t h_X(s)\,ds} \qquad \text{for } X \in \{U, D\}.$$

The bound on the downtime, or accumulated downtime, if this is considered, is denoted by the random variable T with CDF $F_T(t)$. If a deterministic downtime b is considered, then $F_T(t)$ is the unit step function at $t = b$. Let K be the bound on the number of repairs, and N a random variable denoting the number of failures until a full system failure. Denote by $F_{\tilde{X}}(s) = \int_0^\infty e^{-st}\,dF_X(t)$, $X \in \{U, D\}$ the Laplace-Stieltjes transform (LST)[1] of the probability distribution function F.

Let us assume a given system needs to process an amount of work w, which is measured in the time it needs to be processed or the number of commands to be executed. If the work requirement is random, we denote it using the random variable W. In a perfect failure free system the time to finish a job of size w equals time w. In a system that fails the job is interrupted upon failure of the system. Then two possible cases are considered: either the job is resumed after repair of the system (*preemptive resume* failure) or the job has to be restarted anew (*preemptive repeat* failure). In any case it will take longer than w time units to finish the job with work requirement w. Let $T(w)$ denote the time needed to complete a work requirement w in a system subject to failure and repair. For the preemptive resume case and unlimited failures the LST of the job completion time has been derived in [113] as

$$F_{\tilde{T}}(t, w) = \frac{(s + \gamma)e^{-(s+\gamma)w}}{s + \gamma(1 - F_{\tilde{D}}(s)(1 - e^{(s+\gamma)w}))} \tag{2.1}$$

Equation (2.1) is a special case of the respective more general formula in [89]. A similar result has been obtained in [63]. The distribution of the completion time of a job with random work requirement W in a system subject to failure and repair (without checkpointing) as given in (2.1) in the transform domain cannot be used for direct computation of the completion time distribution. But its expectation can be computed using the relationship $E(T(w)) = \frac{-\partial F_{\tilde{T}}(t,w)}{\partial s}\big|_{s=0}$

$$E(T(w)) = \left(\frac{1}{\gamma} + E(D)\right)(e^{\gamma w} - 1). \tag{2.2}$$

It is interesting to observe from (2.2) (and pointed out in [113, 88, 37]) that the time needed to complete the work requirement w, $E(T(w))$ grows exponentially with the work requirement, as shown in Fig. 2.2 for a failure rate of $\gamma = 0.01$ and mean downtime of $E(D) = 0.1$ time units. Repairable systems using a combination of the different types of preemption are a generalised form of the model above. Job completion time in those systems, represented as a semi-Markov model is considered in [88] in very general form.

---

[1] See appendix C.3 for properties of the Laplace and the Laplace-Stieltjes transform

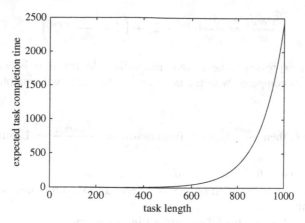

Fig. 2.2 Expected task completion time

For the special case of exponentially distributed time between failures U, or failure rate γ and given work requirement w the probability that the task can be finished is given by the probability that an up period of the system is longer than the task length [16]:

$$\Pr\{U \geq w\} = e^{-\gamma w}. \tag{2.3}$$

After each failure the task must be started again from the beginning, so the assumed failure mode is preemptive repeat.

The mean number of runs needed to complete a task of length w in a system with failure rate γ increases exponentially with the task length and is given by

$$M = e^{\gamma w}. \tag{2.4}$$

The average duration of all runs is [16]

$$T_{\text{average}} = \frac{1}{\gamma}\left(1 - (1 + \gamma w)e^{-\gamma w}\right)\left(1 - e^{-\gamma w}\right) + we^{-\gamma w}. \tag{2.5}$$

Obviously, the higher the failure rate, the shorter the average run length.

The total run time needed to complete one execution of length w is therefore

$$\widehat{T}_{\text{average}} = M \cdot T_{\text{average}} =$$
$$= \frac{1}{\gamma}\left(1 - (1 + \gamma w)e^{-\gamma w}\right)\left(e^{\gamma w} - 1\right) + w \tag{2.6}$$

or equivalently

$$\frac{\widehat{T}_{average}}{w} = \frac{1}{\gamma w}\left(e^{\gamma w} - 2\right) + \left(1 + \frac{1}{\gamma w}e^{-\gamma w}\right). \tag{2.7}$$

Equation (2.7) expresses some system properties. As the time between failures becomes long as compared with the task length most runs will complete the task, i.e.

$$As \frac{w}{1/\gamma} \to 0, \text{ then } \frac{T_{average}}{w} \to 1 \text{ from below and } \frac{\widehat{T}_{average}}{w} \to 1 \text{ from above.}$$

The few runs that still fail have runtime shorter than w, therefore the first limit holds. Since most runs succeed on the long-term average little time is wasted and the second limit holds.

Furthermore, there are the following limiting worst cases.

$$As \frac{w}{1/\gamma} \to \infty, \text{ then} \frac{T_{average}}{w} \to \gamma w \text{and} \frac{\widehat{T}_{average}}{w} \to \infty$$

If, on the other hand, the expected time between failures becomes short as compared with the task length, most runs fail and only very few complete. The average duration of runs lasts until the occurrence of a failure $1/\gamma$ and the number of runs needed to complete the task grows indefinitely.

Under the assumption that failures occur in a Poisson process at rate γ, they do not happen during recovery and are detected immediately, the repair time D, the task length w and the failure rate γ are constants, an appealing expression for the expected task completion time has been derived in [37]:

$$E[T(w)] = \left(D + \frac{1}{\gamma}\right)\left(e^{\gamma w} - 1\right). \tag{2.8}$$

Note that also in this special case the expected task completion time is an exponential function of the task length w.

The assumption of exponentially distributed time to failure is in many cases not appropriate. Because of its ability to model aging, especially in reliability analysis, the Weibull distribution is often used. In [93] the time needed to complete a job of fixed length in a system with preemptive repeat failure mode has been studied. It was shown that if the time to failure is exponentially distributed the mean time needed to finish the task is underestimated if the appropriate failure time distribution would be the Weibull distribution with shape parameter greater than one. The expected time to complete a task of length w is given as[2]

---

[2] It should be mentioned that an implementation of the given equation does not show the property claimed in [93].

$$E[V(w)] = w + \frac{1}{1 - F_U(w)} \left(\int_0^w x f_U(x)\, dx + F_U(w) E[U] \right)$$

This equation can be used to derive a bound on the mean time $E[V(w)]$ needed to complete a task of length w, which is

$$w \leq E[V(w)] \leq w + \frac{E[U] + E[D]}{1 - F_U(w)}.$$

The lower bound is obvious, since completing the job at least needs its length. The upper bound can be interpreted as follows. If most of the probability mass of $f_U(.)$ concentrates in the interval $[0, w)$, then $F_U(w) \approx 1$ and then it is very likely that the up period is not long enough to finish the task and the mean time to complete the task becomes infinitely large. If, on the other hand, most of the probability mass of $f_U(.)$ concentrates in the interval $[w, \infty)$, then $F_U(w) \approx 0$ and most uptimes are longer than w and therefore the mean time to complete the task will be close to the task length.

Going back to the general treatments, clearly, on the average and if failure and repair distribution are continuous and real valued, with preemptive resume failures all jobs will eventually be completed, while with preemptive repeat failures in the worst case the system will have a time to failure that is almost always shorter than the job completion time and hence the job will never be finished.

We can furthermore distinguish systems by the amount of repair they allow for. We consider three cases:

- limited repair time. For instance in real-time systems a job must be finished before a deadline, allowing only for a strictly bounded repair time [60].
- limited accumulated repair time. Availability guarantees of network providers promise a bounded accumulated downtime per time, e.g. per month.
- limited number of repairs. Economic considerations might limit the number of failures a system should experience until it is declared broken.

The analysis carried out in this subsection is extended in [89] by using a general system state model, instead of the simple up-down model we use here. In [89] the system states are described using a semi-Markov process with possibly infinite state space. Subsets of the state space can apply each of the three failure modes, preemptive resume, preemptive repeat identical and preemptive repeat different. If not stated otherwise a preemptive repeat failure indicates a preemptive repeat different one. After a failure and restart the work requirement is again randomly drawn from a distribution. Note that for deterministic as well as exponentially distributed work requirement the distinction between preemptive repeat identical and preemptive repeat different failures does not exist. In this section the two remaining failure modes, i.e. preemptive resume and preemptive repeat failure mode are treated separately. We discuss here a special case of the treatment in [89].

This section will be organised according to the three ways in which repair is limited. Furthermore, only the probability of task completion until a given time is computed, not the full distribution of task completion time.

2.1 Bounded Downtime

In real-time systems a bound on the downtime is especially relevant to define system failure. Complete system failure occurs when the downtime of a system for the first time exceeds a given maximum value. Control systems have to be switched off if the control process remains inactive too long. In this section the work in [60] is reported, as this is the only reference concerned with the situation of bounded individual downtime. The bound B on the downtime is assumed to be deterministic with value b.

Since the length of downtimes is an iid random variable the number of downtimes (N) until the first downtime longer than b and consequently until system failure has a geometric distribution with parameter $F_D(b)$, the probability a downtime is at most of length b,

$$\Pr\{N = n\} = (1 - F_D(b))(F_D(b))^{n-1}, \qquad n = 1, 2, \ldots. \tag{2.9}$$

Then the mean number of downtimes is

$$E[N] = \frac{1}{1 - F_D(b)}. \tag{2.10}$$

2.1.1 System Lifetime

Since the system fails when the n-th downtime takes longer than b the system lifetime consists of the sum of n uptimes, $n - 1$ downtimes and a final downtime that equals b. Hence conditioned on $N = n$ downtimes the system lifetime X_n can be expressed as

$$X_n = nU + (n - 1)D_{<b} + B \tag{2.11}$$

where $D_{<b}$ is the random variable of downtimes less than b. The random variable X_n has CDF $F_{X_n}(t)$ and corresponding Laplace-Stieltjes transform (LST) $F_{\widetilde{X_n}}(s)$

$$F_{\widetilde{X_n}}(s) = (F_{\widetilde{U}}(s))^n (F_{\widetilde{D_{<b}}}(s))^{n-1} e^{-sb} \tag{2.12}$$

and

$$F_{\widetilde{D_{<b}}}(s) = \frac{\int_0^b e^{-sh} \, dF_D(h)}{F_D(b)}. \tag{2.13}$$

Using (2.9) the condition in (2.12) can be removed and the LST of the system lifetime becomes

$$F^\sim(s) = \sum_{n=1}^{\infty}(1 - F_C(b))(F_D(b))^{n-1}\, F_{\tilde{X}_n}(s)$$

$$= \frac{(1 - F_C(b))F_{\tilde{U}}(s)\, e^{-sb}}{1 - F_C(b)F_{\tilde{U}}(s)F_{\tilde{D}_{<b}}(s)}. \tag{2.14}$$

Numerical inversion of (2.14) yields the distribution of the system lifetime. The expected system lifetime is computed as expectation of (2.11) and removing the condition on n. The expected lifetime evaluates to

$$\mathrm{E}[X] = b + \frac{E\{U\} + F_D(b)E\{D_{<b}\}}{1 - F_D(b)} \tag{2.15}$$

where

$$\mathrm{E}[D_{<b}] = \frac{\int_0^b h\, dF_D(h)}{F_D(b)}.$$

For an exponentially distributed downtime with repair rate η the mean system lifetime is obtained by evaluating (2.15) as

$$\mathrm{E}[X] = \left(\mathrm{E}[U] + \frac{1}{\eta}\right)e^{\eta b} - \frac{1}{\eta}.$$

The expected system lifetime for limited downtime is shown in Fig. 2.3 in comparison with the the expected system lifetime for limited cumulative downtime and limited number of repairs. For bounded downtime the expected system lifetime increases exponentially with the repair rate of the system, shown as the straight line in the plot with logarithmic scale. The higher the repair rate, the shorter are the system downtimes and the higher is the probability that the system will live long until it experiences a downtime longer than b for the first time. We set b to 3.0, $\gamma = 1$ and η ranges from 0.0 to 3.0. In most systems typical downtimes are much shorter than the uptimes. A relation of $\nu = 10..100 * \gamma$ is very common. This range is not shown in Fig. 2.3 for better visibility of all curves. All curves can easily be extended to obtain realistic numbers. Note that the picture will be very different for distributions with a heavy tail or for e.g. the uniform distribution.

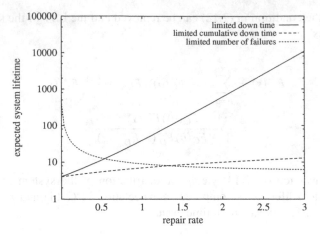

Fig. 2.3 Expected system lifetime

2.1.2 Cumulative Uptime

To obtain the probability of completing a task the system lifetime is not needed. Instead the cumulative uptime is used, which is computed similarly to the system lifetime.

The cumulative uptime is the sum of n uptimes, under the condition of observing exactly n downtimes.

$$Y_n = nU. \tag{2.16}$$

Similar as for the system lifetime, taking the LST and removing the conditioning on n gives

$$F_{\tilde{Y}}(s) = \frac{(1 - F_D(b))F_{\tilde{U}}(s)}{1 - F_D(b)F_{\tilde{U}}(s)}. \tag{2.17}$$

In general (2.17) must be inverted numerically. For exponentially distributed up- and downtimes with rates γ and ν the distribution of the cumulative uptime evaluates to

$$F_Y(t) = 1 - e^{-\gamma t e^{\eta b}} \tag{2.18}$$

which is an exponential distribution with rate $\gamma e^{-\nu b}$. The mean cumulative uptime is obtained as the expectation of the LST of (2.16) with removing the condition on n:

$$E\{Y\} = \frac{E\{U\}}{1 - F_D(b)}. \tag{2.19}$$

2.1.3 Probability of Task Completion

To finish a task with work requirement w the system needs a cumulative uptime of length w. We are interested in the probability $\Pr\{w\}$ the task will be finished before the system fails. Here we have to distinguish between the two failure modes, *preemptive resume* and *preemptive repeat*. In the resume mode we obtain

$$\Pr\{w|N = n\} = \Pr\left\{\sum_{i=1}^{n} U_i > w\right\} =$$
$$= 1 - \Pr\left\{\sum_{i=1}^{n} U_i \leq w\right\} =$$
$$= 1 - F_U^{(n)}(w). \tag{2.20}$$

Unconditioned, the probability of task completion can be expressed as the probability of observing accumulated system lifetime that is at least w.

$$\Pr\{w\} = 1 - F_Y(w), \tag{2.21}$$

which for an exponentially distributed system lifetime evaluates to

$$\Pr\{w\} = e^{-\gamma w e^{-\eta b}}. \tag{2.22}$$

In the preemptive repeat failure mode the task completes if the length of at least one out of n uptimes is w or more.

$$\Pr\{w|N = n\} = 1 - (F_U(w))^n \tag{2.23}$$

The condition is removed using (2.9)

$$\Pr\{w\} = \frac{1 - F_U(w)}{1 - F_D(b)F_U(w)}, \tag{2.24}$$

which for exponentially distributed up- and downtimes evaluates to

$$\Pr\{w\} = \frac{1}{1 - (1 - e^{\gamma w})e^{-\eta b}}. \tag{2.25}$$

For both failure modes, as $b \to 0$, $P(w) \to e^{\gamma w}$, the probability of completing the task within one uptime increases exponentially, whereas as $b \to \infty$, $P(w) \to 1$. If infinitely long down times can be tolerated the task will eventually be finished with certainty. Figure 2.4 shows the probability of task completion for preemptive resume and preemptive repeat failure mode as a function of vb for $\gamma w = 3$.

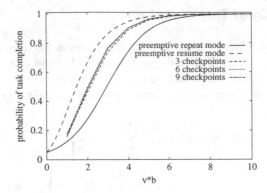

Fig. 2.4 Probability of task completion for bounded downtime

Following the intuition, for exponentially distributed up- and down times the preemptive resume failure mode leads to higher probability of task completion.

For the time to failure the exponential distribution is perhaps not the best choice, but a reasonably good one. For the repair time, one might rather chose a uniform distribution, since for all repair actions there is a lower time limit and most repair takes not too long. The Laplace transform, however, is not invertible for all probability distributions and its inversion can be a painful task. This limits the general applicability of models in Laplace transform domain.

2.2 Bounded Accumulated Downtime

As a modification of the assumptions in the previous section, in this section we consider bounded accumulated downtime. Again, a system may be up or down, but once the accumulated downtime exceeds a given bound the system is considered failed. Such systems can be found where e.g. contracts or service level agreements guarantee a bound on the downtime per time interval. Such contracts are used in highly reliable fault tolerant information systems or in telecommunication networks, etc. The work in [114] extends [60] further by considering the work requirement W and the bound on the cumulative downtime B to be random variables.

When modelling a system of redundant components the cumulative downtime is well represented by a random variable rather than a deterministic value. A backup server takes over operation each time the primary server fails. The downtime of the primary server therefore reduces the lifetime of the backup server. The system eventually fails completely when the backup server fails. Hence the time to failure of the backup server is the random variable that bounds the cumulative downtime of the system.

Let us again, as in [60], formulate expressions for the number of downtimes. The probability of observing more than n downtimes until system failure is equivalent to the probability that the sum of n downtimes is less than the deterministic bound b.

$$Pr\{N > n\} = F_D^{(n)}(b), \qquad n = 0, 1, \ldots,$$

where $F_D^{(n)}(b)$ is the n-fold convolution of the downtime distribution. Therefore

$$Pr\{N = n\} = F_D^{(n-1)}(b) - F_D^{(n)}(b), \qquad n = 1, 2, \ldots, \qquad (2.26)$$

and the expected number of downtimes equals

$$E\{N\} = 1 + \sum_{n=1}^{\infty} F_D^{(n)}(b). \qquad (2.27)$$

As an example, for exponentially distributed downtimes with parameter v, the probability of exactly n downtimes until system failure is

$$Pr\{N = n\} = \frac{(vb)^{n-1}e^{-vb}}{(n-1)!}, \qquad n = 1, 2, \ldots, \qquad (2.28)$$

with expectation

$$E\{N\} = 1 + vb.$$

2.2.1 System Lifetime

The system lifetime is determined by investigating the remaining potential down-time at the moment where an up period begins. $X(t)$ is the remaining system lifetime at the start of an uptime given a cumulative downtime of already t, such that the remaining accumulated downtime is $b - t$. The system lifetime will now consist of the just started uptime plus, potentially, another uptime, if the downtime in between is not too long. We have to distinguish two cases. Either the following downtime h will be at least as long as t, then the system lifetime will end after this uptime and time t of the next downtime. If, on the other hand, the next downtime is shorter than t, the system lifetime will be the sum of this uptime, the next downtime and then another remaining system lifetime with a given remaining bound $t - h$. In other words,

$$X(t) = \begin{cases} U + t, & \text{with probability } 1 - F_D(t) \\ U + h + X(t-h), & \text{with probability } dF_D(h), 0 < h < t. \end{cases}$$

Taking first the LST and then the Laplace transform with respect to t gives after some manipulation [60, 114]

$$F_X^{\sim*}(s, v) = \frac{(1 - F_D^{\sim}(s+v))F_U^{\sim}(s)}{(s+v)(1 - F_U^{\sim}(s)F_D^{\sim}(s+v))} \qquad (2.29)$$

In [60], for exponentially distributed up- and downtimes the above equation has be inverted with respect to v

$$F_{\widetilde{X}}(s, b) = \frac{\gamma}{s + \gamma} \exp\left(-\left(s + v - \frac{\gamma v}{s + \gamma}\right)b\right),$$

which must then be further inverted numerically.

If the accumulated downtime is an exponentially distributed random variable T with parameter β instead of a constant b, transformation with respect to b is not necessary and (2.29) simplifies to [114]

$$F_{\widetilde{X}}(s) = \frac{\beta F_{\widetilde{U}}(s)(1 - F_{\widetilde{D}}(s + \beta))}{(s + \beta)(1 - F_{\widetilde{U}}(s)F_{\widetilde{D}}(s + \beta))}. \tag{2.30}$$

Because of the memoryless property of the exponential distribution, for an exponentially distributed accumulated bound on the downtime the same result holds as for an exponentially distributed bound on a single downtime.

The mean system lifetime still can be computed in a straight forward manner. It consists of a total downtime of length b and the expected number of uptimes $E\{N\}$, multiplied with the expected length of an uptime $E\{U\}$

$$E\{X\} = B + E\{N\}E\{U\},$$

and $E\{N\}$ is given in (2.27). The expected system lifetime, assuming exponential up- and downtimes is shown in Fig. 2.3 on p. 20, where $b = 3.0$, the system failure rate $\gamma = 1$ and the repair rate v varies from 0.0 to 3.0. The system lifetime in systems with bounded accumulated downtime is a straight line with respect to the mean downtime (or its inverse, the repair rate). Since Fig. 2.3 has a logarithmic scale the system lifetime has shape of the logarithm function. The higher the repair rate the shorter are the down periods, and, consequently, the longer are the relative uptimes per time interval. It takes longer to accumulate a given downtime as the expected downtime becomes shorter. Obviously, accumulating a down time of $b = 3.0$ happens for most up- and downtime distributions much faster than observing a single downtime of length $b = 3.0$. A system which can tolerate all downtimes shorter than b can in fact tolerate much more downtime than a system with a bound on the accumulated downtime and has therefore a longer expected lifetime.

2.2.2 Cumulative Uptime

The cumulative uptime is the sum of the uptimes until system failure, i.e. the difference between the system lifetime and the sum of the downtimes until system failure. The sum of the downtimes until system failure is by definition the bound on the accumulated downtime b and therefore the cumulative uptime is computed as

$$Y = X - b$$

The LST of the distribution of the cumulative uptime is obtained by conditioning on the number of uptimes as given in (2.26)

$$F_{\widetilde{Y}}(s) = \sum_{n=1}^{\infty} Pr\{N = n\}(F_{\widetilde{U}}(s))^n.$$ (2.31)

For exponentially distributed up- and downtimes using (2.28) Eq. (2.31) evaluates to

$$F_{\widetilde{Y}}(s) = \frac{\gamma}{s+\gamma} \exp\left(-\frac{vbs}{s+\gamma}\right),$$ (2.32)

which still needs to be inverse transformed.

2.2.3 Probability of Task Completion

For computing the probability that a task with work requirement w can be completed before failure of the system the two failure modes (preemptive resume and preemptive repeat) have to be distinguished.

In the resume failure mode the probability of finishing a given task equals the probability of having a cumulative uptime longer than w, which is

$$Pr\{w\} = 1 - F_Y(w).$$

Unfortunately, F_Y can only be expressed in the transform domain. Even for exponentially distributed up- and downtimes the inverse transformation must be carried out numerically. This is a severe limitation of transforms as a modelling tool. In this text we therefore do not plot the probability of task completion in the resume mode.

For preemptive repeat failures the probability of task completion is the probability of having at least one uptime of length w, which is obtained by conditioning on the number of downtimes until system failure

$$Pr\{w\} = 1 - \sum_{n=1}^{\infty} Pr\{N = n\}(F_U(w))^n.$$ (2.33)

$Pr\{N = n\}$ again is given in (2.26). For the preemptive repeat case a closed form solution is available. Substituting (2.28) into (2.33) for exponentially distributed up- and downtimes at failure rate γ and repair rate v leads to

$$Pr\{w\} = 1 - (1 - e^{-\gamma w})e^{-vb\,e^{-\gamma w}}.$$ (2.34)

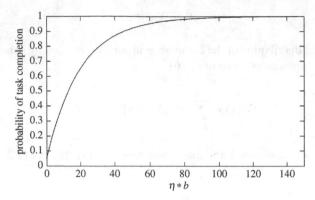

Fig. 2.5 Task reliability for bounded accumulated downtime (preemptive repeat failure mode, $\gamma w = 3$)

The graph of the probability of task completion for bounded cumulative downtime in the preemptive repeat case is shown in Fig. 2.5. It is equal to the task reliability.

Again using the deterministic bound b on the cumulative downtime, a task of deterministic length w and preemptive resume failure mode, in [114] the task completion time $C(w, b)$ is considered a random variable with probability distribution function $P(t, w, b) = P(C(w, b) \le t)$. Because of the bounded downtime, in general $P(t, w, b)$ is a defective distribution. Its LST is denoted

$$P^{\sim}(s, w, b) = \int_0^\infty e^{-st} dP(t, w, b)$$

The probability of completing the task, the task reliability can alternatively be computed as

$$\eta(w, b) = Pr\{C(w, b) < \infty\} = P^{\sim}(0, w, b). \tag{2.35}$$

The main result in [114] is the formulation of the LST with respect to t and the Laplace transform with respect to both w and b, which are

$$P^{\sim*}(s, u, b) = \int_0^\infty e^{-uw} P(s, w, b) dw$$

and

$$P^{\sim**}(s, u, v) = \int_0^\infty e^{-bv} P(s, u, b) db$$

The triple transform evaluates to

$$P^{\sim **}(s, u, v) = \frac{1 - F_U^{\sim}(s + u)}{v(s + u)(1 - F_D^{\sim}(s + v)F_U^{\sim}(s + u))}. \qquad (2.36)$$

The distribution of the task completion time is obtained by inverting (2.36) with respect to all variables s, u and v. In general numerical inversion is necessary, in some special cases analytical inversion might be possible. Computing the task reliability requires only two (rather than three) inversions.

If the bound on the accumulated downtime is an exponentially distributed random variable T with parameter β and the work requirement is an exponentially distributed random variable W with parameter α then the transformations with respect to b and w can be avoided and the task completion time distribution in LST domain equals

$$P^{\sim}(s) = \frac{\alpha(1 - F_U^{\sim}(s + \alpha))}{(s + \alpha)(1 - F_D^{\sim}(s + \beta)F_U^{\sim}(s + \alpha))}. \qquad (2.37)$$

Whether (2.37) is inverted numerically or analytically depends on the nature of the LSTs of the distributions of the up- and downtimes. The task reliability is computed as $P^{\sim}(0)$ and evaluates to

$$\eta(\alpha, \beta) = \frac{1 - F_U^{\sim}(\alpha)}{1 - F_D^{\sim}(\beta)F_U^{\sim}(\alpha)}.$$

From the task reliability rather obvious relationships between the value of the bound and the cumulative downtimes can be observed. If no bound on the cumulative downtime exists ($\beta = 0$) the task will always complete and $\eta(\alpha, \beta) = 1$. If on the other hand the bound on the cumulative downtime is zero ($\beta \to \infty$), the system fails at its first downtime, then $\eta(\alpha, \beta) = 1 - F_U^{\sim}(\alpha)$ and the task must complete within the first and only uptime.

In preemptive repeat identical failure mode, where after each failure the task is restarted anew and each retry has exactly the same work requirement similar analysis leads to a task reliability of

$$\eta(w) = \frac{1 - F_U(w)}{1 - F_D^{\sim}(\beta)F_U(w)}. \qquad (2.38)$$

The probability of task completion depends on the work requirement w as well as on the length of the up- and downtimes, where only the distribution of the downtimes is given as LST transform. For exponentially distributed up- and downtimes, again with parameters γ and η, respectively, the task reliability can be expressed explicitly depending on the work requirement and the cumulative bound on the downtime and equals the probability of completing the task as given in (2.34)

$$\eta(w, b) = 1 - (1 - e^{-\gamma w})e^{-\nu b e^{-\gamma w}}. \qquad (2.39)$$

It should be noted that at least for exponentially distributed up- and downtimes the task reliability is equal to the probability of task completion as defined in [60]. When comparing Fig. 2.4 on p. 22 and 2.5 the impact of the two failure modes is not obvious. As could be seen in Fig. 2.3 a system with bounded accumulated downtime has considerably shorter system lifetime than a system with bounded individual downtime. On the other hand we expect the preemptive resume failure mode to lead in most cases to faster task completion than the preemptive repeat failure mode. Figure 2.5 suggests that the impact of the failure mode is much stronger than the impact of the bound on the downtime. But this observation remains unproven.

In [114] phase-type distributions for the up- and downtimes, the bound on the downtime and the work requirement are used.

For preemptive repeat different failures a new work requirement is sampled each time a task is restarted after a failure. Obviously, for deterministic work requirement the task reliability is the same with preemptive repeat different failures as with preemptive repeat identical failures.

2.3 Bounded Number of Failures

A bounded number of failures can occur naturally in systems consisting of a set of redundant components, where each failure is the non-repairable failure of one component and once all components have failed, the system has failed.

Since the bound on the number of downtimes is K the system fails upon the $K - th$ transition to the down state and the number of down times equals K with probability one. The last downtime has length zero, since the system immediately moves to the failed state.

2.3.1 System Lifetime

The system lifetime consists of K uptimes and $K - 1$ downtimes, i.e.

$$X = K \cdot U + (K - 1) \cdot D.$$

The mean system lifetime is obtained by taking the expectation of the system lifetime

$$E\{X\} = K \cdot E\{U\} + (K - 1) \cdot E\{D\}.$$

2.3.2 Cumulative Uptime

The cumulative uptime is the sum of K uptimes

$$Y = K U$$

with distribution function

$$F_Y(y) = F_U^{(K)}(y) \qquad y \geq 0 \qquad\qquad (2.40)$$

and expectation

$$E\{Y\} = K * E\{U\}.$$

2.3.3 Probability of Task Completion

In the preemptive resume failure mode the probability of task completion equals the probability that the cumulative uptime before system failure is at least as long as the work requirement of the task. Using (2.40) the probability of task completion evaluates to

$$P(w) = 1 - F_U^{(K)}(w).$$

For exponentially distributed uptimes the $K-$fold convolution evaluates to a $K-$phase Erlang distribution and the probability of task completion becomes

$$P(w) = e^{-\gamma w} \sum_{j=0}^{K-1} \frac{(\gamma w)^j}{j!}.$$

In preemptive repeat failure mode a task completes if at least one uptime is at least as long as the work requires

$$P(w) = 1 - (F_U(w))^K.$$

For exponentially distributed uptimes the above probability is simply computed as

$$P(w) = 1 - (1 - e^{-\gamma w})^K.$$

Figures 2.6 and 2.7 show the probability of task completion again for exponentially distributed up- and downtimes and for both failure modes versus the number of tolerated downtimes.

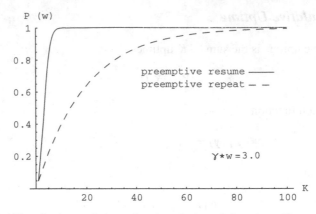

Fig. 2.6 Probability of task completion as function of tolerated downtimes K

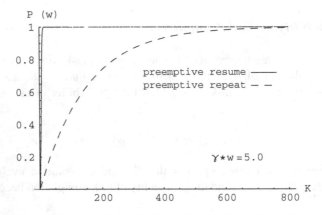

Fig. 2.7 Probability of task completion as function of tolerated downtimes K

At the given failure rate and work load in preemptive resume mode a task completes within few uptimes with high probability.

If the work requirement or the failure rate increases in preemptive resume failure mode a job still finishes quickly, while in preemptive repeat mode it takes considerably more uptime to complete the task with high probability. The shorter the uptimes, or the longer the given task takes, the lower the probability of observing soon an uptime that is long enough to finish the task.

To summarise, in this section we have seen the impact of system downtimes on system performance. Obvious but still interesting observations can be made by looking at the limiting case of a system that does not fail, and revisiting the considered metrics. In a system that does not fail, system lifetime is infinite and the probability of task completion is always one. This limiting result is obtained from all respective equations when the failure rate $\gamma = 0$. Furthermore, neither the characteristic of the bound on the downtime nor the failure mode (preemptive resume or preemptive repeat) have any influence on system lifetime or task completion probability in this special case.

An algorithm for computing the distribution of the job completion time in degradable fault-tolerant systems is presented in [88].

The relations between the bounds on the tolerated system downtime and the system lifetime or probability of task completion formulated in this chapter are intuitively self-evident. Important are their mathematical formulations. The purpose of restart, rejuvenation and checkpointing strategies will be to improve on system lifetime and task completion probability.

While the restart method corresponds to intentionally triggered preemptive repeat different failures the preemptive resume failure mode constitutes the best case obtained with checkpointing. In the models studied so far there is no cost or time penalty involved with restart after a failure and work is continued exactly where it was interrupted.

The preemptive repeat failure mode can also be related with checkpointing where it comes close to the worst case of a checkpointing scheme. With each down period the task needs to be restarted from scratch. As we will see checkpointing could make matters worse, if induced overhead is considered as well.

Part II
Restart

Chapter 3
Applicability Analysis of Restart

The investigation of pure restart as a means to improve performance of computer systems is motivated by an observation known to most Internet users: when a page takes too long to load in a web browser clicking the reload button in many cases helps. In [86] the technical background of this fact is discussed in detail, explaining how clicking the reload button 'overrules' the TCP retransmission timer, potentially improving the overall download time.

In this chapter we will first look at systems, protocols or algorithms where restart occurs or is being applied intentionally. In the second part of this chapter the restart problem is formulated as a stochastic problem. Then stochastic criteria for the successful application of the restart method can be derived.

3.1 Applications of Restart

The term restart applies to job, or task processing systems as well as to transaction processing systems. In all those a job or transaction is issued and usually completes after a certain time. Completion can be defined in different ways. In case of a computation it can be determined by a result being returned. In case of a web service or data base request it can be a message being returned. If a job or transaction does not complete within a certain time, it is re-issued. In some cases the old instance can be aborted, in others it cannot. Furthermore, in some restart scenarios a task is restarted in exactly the identical configuration or parameterisation in other scenarios the restarted task may have a new set of parameters perhaps even a new work requirement.

3.1.1 Randomised Algorithms

The term *Randomised algorithms* is used for algorithms that are not deterministic in their execution, either with respect to runtime, or with respect to the obtained result. This is because some parameters of the algorithm are chosen randomly. Examples of randomised algorithms are those theorem proving algorithms that search the state

K. Wolter, *Stochastic Models for Fault Tolerance*,
DOI 10.1007/978-3-642-11257-7_3, © Springer-Verlag Berlin Heidelberg 2010

space and do so in a random way. If the algorithm does not terminate within a defined number of steps it starts again from the initial state, again randomly choosing a path. Another well-known example is the sorting algorithm *Quicksort*, which speeds up sorting a list by first permuting the entries randomly. For a randomised algorithm that searches a tree to prove a theorem a strategy defines the sequence of numbers of steps it takes in each trial.

Two important types of randomised algorithms exist. *Monte Carlo* algorithms provide the correct result in most cases, but also have a small probability of computing a wrong result. A Monte Carlo algorithm usually completes fast, even when solving a difficult, multi-dimensional problem but the result will be wrong with small probability. Monte Carlo algorithms are used to solve problems in quantum computing, complex models in physics or risk models where not only the result but also the input has some uncertainty.

The second class of randomised algorithms is called *Las Vegas* algorithms and consists of those that will always provide a correct result but have a random and sometimes very long runtime [99, 5]. Note that Las Vegas algorithms can be transformed into Monte Carlo algorithms by just stopping the computation after a fixed time and providing an arbitrary (possibly wrong) result in case the algorithm has not computed the correct answer yet [108]. The randomised Quicksort as well as theorem provers searching a state space are Las Vegas algorithms.

Search in trees is a solution to many problems in artificial intelligence. In [59] complete, systematic, backtrack-style search in tree structures has been identified as one of the main applications of randomised algorithms. Tree structures are not only used for theorem proving but also in circuit design, logistics scheduling and timetabling. Search in trees often has a heavily-tailed cost function. In [59] cut-off values for restart have been determined experimentally resulting in much lower mean problem solution time and hence lower cost. Improved strategies for finding a good cut-off value (restart time) are discussed in [67] whereas the tree-structures that lead to heavy-tailed cost functions for search algorithms are further classified in [171].

A more theoretically profound investigation is carried out in [99]. The runtime, measured in a discrete number of computation steps, of Las Vegas algorithms can be modelled as a random variable with some probability distribution. The completion time of the algorithm is a random number drawn from that distribution. Since the runtime of the algorithm depends on the algorithms input, which itself is randomly chosen, it can be worthwhile to abort the computation at some point and restart.

Randomised algorithms can also be run with several replicas simultaneously. As soon as one replica obtains the correct result all replicas are stopped. Strategies for how long to run each replica and when to restart have been investigated in [99]. There are strategies that require knowledge of the probability distribution for the runtime of the algorithm for a given input. Others can do with only the mean completion time, or completely without any knowledge of the completion time distribution.

3.1.2 Optimal Restart Time for a Randomised Algorithm

Let us first assume the runtime distribution of a randomised algorithm is known. The runtime of a randomised algorithm is measured by means of a discrete random variable, counting the number of computation steps needed until completion. In later sections in this chapter completion time or waiting time is modelled using a continuous random variable.

Let A be an algorithm running a strategy S on input x. Then $A_S(x)$ finishes after exactly t steps with probability $p(t)$. Let $F(t)$ be the cumulative discrete probability distribution and let furthermore the random variable T denote the runtime of the algorithm. We are interested in the expected runtime $E(T_S)$ using strategy S. A strategy defines the number of steps that e.g. a search algorithm is executed and possibly what branch to take when there is a choice. A more complex search strategy defines several sequences of steps to be executed consecutively.

Let $S = (t_1, t_2, t_3, \ldots)$ be an execution strategy consisting in several attempts each of length t_i and let $S_1 = (t_2, t_3, \ldots)$ be that same strategy without the first trial. $E(T)$ denotes the expected running time using strategy S, whereas $E(T_1)$ is the expected running time when using S_1. Then we can formulate a recursion for the expected runtime of the algorithm: either it completes in the first trial, with the respective expected runtime, or after having executed t_1 steps it is aborted and restarted, needing those t_1 steps plus the expected runtime thereafter. In mathematical terms that is

$$E(T) = \sum_{i=1}^{t_1} i \cdot p(i) + (1 - F(t_1))(t_1 + E(T_1)). \tag{3.1}$$

If all strategies are the same, i.e. $t_1 = t_2 = \ldots = t$ and the same number of steps is computed in every trial then Eq. (3.1) simplifies to

$$E(T_t) = \frac{1}{F(t)} \sum_{i=1}^{t} i \cdot p(i) + t \frac{(1 - F(t))}{F(t)}. \tag{3.2}$$

In [99] the equivalence

$$\sum_{i=1}^{t} i \cdot p(i) = t F(t) - \sum_{i=1}^{t-1} F(i) \tag{3.3}$$

is used to derive the following expression for the expected runtime of an algorithm executing the optimal strategy

$$l(t) = E(T_t) = \frac{1}{F(t)} \sum_{i=1}^{t} i \cdot p(i) + t \, \frac{(1 - F(t))}{F(t)}$$

$$= \frac{1}{F(t)} \left(t \, F(t) - \sum_{i=1}^{t-1} F(i) \right) + t \, \frac{(1 - F(t))}{F(t)}$$

$$= \frac{1}{F(t)} \left(t - \sum_{i=1}^{t-1} F(i) \right). \tag{3.4}$$

For a proof of (3.3) see Appendix 10.

Luby et. al. [99] furthermore show that the strategy with runs of equal length $S^* = (t, t, \ldots)$ is optimal for every probability distribution p, since no other strategy will give a lower expected running time. The runtime t can be bounded from above by $E[p]$, the expectation of the distribution f, but no explicit formulation that could help to determine t is given.

If the runtime distribution of an algorithm is unknown the expected runtime can be bounded for all possible distributions. In [99] no bound is provided for the optimal strategy $(t, t, \ldots)$, instead a *universal* strategy is used that has a structure based on powers of two

$$S_{\text{univ}} = (1, 1, 2, 1, 1, 2, 4, 1, 1, 2, 1, 1, 2, 4, 8, 1, \ldots) \tag{3.5}$$

where each time a pair of a given run length has been completed a run of twice that length follows.

Formally the universal strategy is defined as

$$t_i = \begin{cases} 2^{k-1}, & \text{if } i = 2^k - 1 \\ t_{i-2^{k-1}+1}, & \text{if } 2^{k-1} \le i < 2^k - 1, \end{cases} \tag{3.6}$$

for $k \in \mathbb{N}$.

The complexity of this strategy can be bounded from above for all possible runtime distributions using the optimal strategy by the following expression in terms of the expected runtime

$$E(T_{\text{univ}}) \le 192 \cdot E(T_t) \, (\log(E(T_t)) + 5). \tag{3.7}$$

As stated in [99] a tighter bound could be derived in a less compact and elegant expression.

Finally, as long as the runtime distribution is unknown any strategy will have at least complexity $E(T_t) \cdot \log E(T_t)$, if $E(T_t)$ is the expected runtime under the optimal strategy. In other words

$$\sup E(T) \geq \frac{1}{8}(E(T_t) \cdot \log E(T_t))$$

as proven in [99].

In practice the runtime distribution typically is not known and it is much easier to estimate the mean runtime. Therefore the mean runtime is used in a pragmatic rule, such as aborting after twice the expected runtime. This will reduce the probability of longer runs with exponential rate (A. Taraz, December 2004, personal communication).

A follow-up application is that of distributed queries using search algorithms that have a random aspect. This has recently been studied in, e.g., [132], and online algorithms have been derived to set the restart time if dependencies between successive tries can be exploited.

3.1.3 Failure Detectors

Fault-tolerant distributed systems are designed to provide reliable and continuous service despite failure of some of their components. A failure detector is a basic building block of such systems. A failure detector is used by a monitoring process to determine whether the monitored process is still up and running. Two types of failure detectors exist. The ping style failure detector and the heartbeat type failure detector.

Using a heartbeat failure detector, the monitored process periodically sends an *alive* message to the monitoring process. A timeout determines the maximum waiting time of the monitoring process until it suspects the monitored process to have failed. While heartbeat failure detectors implement the push technology, a ping-style failure detector employs a pull mechanism. When using a ping failure detector the monitoring process periodically sends messages to the monitored process. If no acknowledgement arrives until the expiry of a timeout the monitoring process suspects the monitored process to have failed.

Of interest is the accuracy of failure detection and the detection time. The accuracy is defined as the probability of the suspicion being correct and the monitored system having in fact failed. The detection time is the period between failure of the system and failure detection by the monitoring system.

A good choice of the timeout value in the monitoring process is essential when optimising both metrics [27, 11]. If the timeout is too small then the monitored system is suspected of having failed too often, while if the timeout is too long detection time increases unnecessarily.

The conclusion that *it is impossible to create a failure detection mechanism with the best accuracy and delay together* [39] corroborates the concept of a tradeoff between timeliness and fairness for the restart model as elaborated in [129]. The restart model as it is discussed in Chap. 4 has been proposed to determine the timeout value of failure detectors in [162, 161]. Even though clock synchronisation poses a serious problem heartbeat failure detectors are considered superior in [11], where, as in [130], different known algorithms such as the TCP algorithm [74] are used for timeout computation.

3.1.4 Congestion Control in TCP

TCP (Transmission Control Protocol) is a protocol operating in networks with varying number of participants connecting different subnets using very different technologies. To assure some quality and speed in data transmission congestion avoidance and congestion control are essential [145].

TCP acknowledges the receipt of every packet that has been sent. A sender can send only a limited number of packets while waiting for open acknowledgements. The accepted number of pending acknowledgements is called the *congestion window size*. As data flows the number of allowed pending acknowledgements is increased, i.e., the congestion window opens up.

If no acknowledgement arrives to the sender within a certain time called the *retransmission timeout* (RTO) the packet is being resent. There can be mainly two reasons for not receiving an acknowledgement: either the packet or its acknowledgement was damaged, or there is congestion on the communication channel and the packet got lost in a queue overflow. Packets have been reported in the not very recent publication [74] to be damaged in less than 1% of lost packets. Today, in wireless networks the packet loss rate can be 25% or more, depending on the network topology. Actual loss rates are rarely released for publication by network providers and therefore very difficult to find. For some backbone networks service-level-agreements (SLAs) exist and measurements are published proving that the network Quality-of-Service (QoS) does not violate the SLA [141]. We found in our experiments of connection setup over the Internet that roughly 0.5% of all Internet connections time out at least once. From this observation one can conclude, that the observed end-to-end loss rate in our experiments was less than 0.5% [128].

TCP can in many cases guarantee data transmission even in the presence of data loss. This is achieved through retransmission of unacknowledged data packets. The performance of TCP depends to a large extent on the choice of the timeout between retransmissions.

Several proposals for the choice of the RTO exist. Most TCP implementations today use the Jacobson-Karn algorithm [74], which bases the RTO for one packet on an estimate of the round-trip-time (RTT) computed from the observed RTT of previous packets.

In each step of the algorithm the RTO is set to

$$RTO = \beta R$$

where β accounts for the variation in RTTs of previous packets [74] and R is an estimate of the RTT computed as

$$R_{\text{new}} = \alpha R_{\text{old}} + (1 - \alpha)M$$

with M being the RTT estimate of the most recently acknowledged packet. The parameter α is a filter typically chosen as $\alpha = 0.9$.

Today's TCP implementations typically use for the first few packet transmissions the following fixed values for their RTO timer: 3, 6, 12, 24, 48, etc. and adjust the RTO according to the proposal in [74] with implementation guidelines formulated in the TCP standard [120] as an estimate of the RTT is obtained. The RTO is in some cases reduced down to 250–300 ms [86]. Starting with 3s in each step the previous value is doubled, obtaining an exponential back-off. It is, however, unclear whether the base modulus of 3 s indeed leads to optimal performance.

In HTTP transactions most users experience the RTO value of TCP as suboptimal and 'work around' it by clicking the *reload* button of their web browser. The question whether there exists an optimal RTO value for HTTP traffic motivated our investigations, that will be discussed in detail in the next chapter.

We carried out experiments measuring the duration of the connection-setup in a TCP connection, which consists of a three-way handshake as depicted in the sequence diagram in Fig. 3.1.

Figure 3.2 shows data sampled for the time needed to perform the three-way handshake forming the connection setup. The sample consists of roughly 230,000 experiments connecting to approximately 200 different hosts world-wide. The horizontal lines clearly show the retries carried out by TCP after expiration of the RTO timer. The timeouts are so long that a delay in an HTTP transfer can be traced back to the expiration of a TCP timer among other potential reasons. Restart of HTTP transactions has to some extent already been investigated in [136]. A generalisation are Internet agents [22, 104]. Internet agents carry out varying tasks, using possibly randomised algorithms, over networks with failures and unpredictable delays, and it may therefore be smart to interrupt and restart an agent's job when a task takes too long to complete.

In a more general sense, however, restart has been around in computing systems since their inception. Timeout schemes that retry an attempt once a threshold has

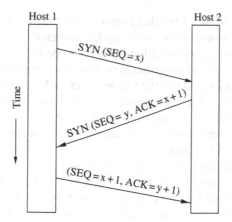

Fig. 3.1 Three-way handshake for connection set-up in TCP

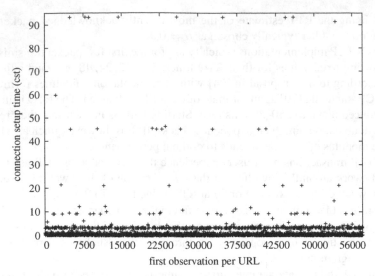

Fig. 3.2 Connection set-up times for the first connection per URL download

been reached, can be seen as restarts. The TCP retransmission timeout is but one example.

In the next sections we will carry out a pure theoretical study of restart. We will investigate necessary and sufficient properties of response times or job completion times that make restarting a beneficial action. We will look at different metrics to be optimised under restart and, furthermore, propose algorithms to determine best times for when to restart.

3.2 Criteria for Successful Restarts

What characteristics do jobs have that benefit from restarts? This question cannot be asked as straight forward as it is formulated, although it turns out that the answer to the different precise questions will in many cases be the same. We have to consider the metric of interest in the question we ask. So, our question becomes: what characteristics do jobs have so that their expected completion time will benefit from restart? We can also ask: what characteristics do jobs have such that their probability of meeting a given deadline will improve under restart?

We limit ourselves here to those two questions, which can be generalised towards optimising quantiles under restart.

The models we set up are valid under two assumptions: (1) we assume that the restart of a task terminates the previous attempt. This is for instance the case when we click the reload button in a web browser: the connection with the server is terminated and a new download attempt is issued [86]. And (2) we then assume that successive tries are statistically independent and identically distributed. This assumption has been found realistic in a measurement study of HTTP Get [128]. It should be noted that both above assumptions have also been found realistic for the use case of downloading web pages by others [104].

In mathematical terms, the problem formulation is as follows. Let the random variable T denote the completion time of a job, with probability distribution $F(t), t \in [0, \infty)$, and probability density function $f(t), t \in [0, \infty)$. Assume τ is a restart time, and introduce the random variable T_τ to denote the completion time when an unbounded number of retries is allowed. That is, a retry takes place periodically, every τ time units or until the deadline has passed (if a deadline is set), which ever comes first. We write $f_\tau(t)$ and $F_\tau(t)$ for the density and distribution of T_τ, respectively.

We will see that the hazard rate function is indicative of whether restart is beneficial irrespective of the considered metric. Fairly general, one can say that if the task completion time distribution has an increasing hazard rate restart should not be applied while for decreasing hazard rate restart is always beneficial.

The hazard rate is commonly used in survival analysis, which is mostly used in medical studies, but also in reliability theory. Survival analysis is concerned with the number of deaths within some time interval and with the number of individuals surviving some point in time. In reliability theory [57] the hazard rate is called failure rate since the failure of the system is its 'death'. As we will see later, preventive maintenance is useful in systems with increasing failure rate (IFR), (as opposed to decreasing failure rate (DFR)) where failures become more frequent as the system ages. Restart, on the contrary, is only beneficial for tasks that are more likely to complete soon and hence have decreasing hazard rate.

Many distributions, however, have a hazard rate which partly increases and partly decreases. We will therefore address the question of when restart should be applied more formally.

3.2.1 When Does Restart Improve the Expected Completion Time?

In general terms, the completion time when starting new must be less than the completion time when not restarting. Assume we are interested in the mean completion time. Under the assumption of independent identically distributed completion time of successive tries, one would restart at time τ when:

$$E[T] < E[T - \tau | T > \tau]. \tag{3.8}$$

The above intuitive reasoning can be made precise, and indeed turns out to be correct. Even stronger, in Sect. 4.3.1.3 we will show that (3.8) is a necessary and sufficient condition for *any* number of restarts to be useful as well, a result that is not necessarily obvious at first hand. The question then becomes, what distributions fulfil requirement (3.8), for at least one value of τ. Generally speaking, first, distributions with heavy tails have the required behaviour. For such distributions, the tail decreases at polynomial pace, leaving considerable probability mass at high values of T. Heavy-tailed and similar distributions commonly arise when studying Internet applications, see for instance [86]. However, also distributions with exponentially decaying tails demonstrate the required behaviour quite often. Considering the three prototypical cases of exponentially decaying distributions [69], we see the following: for hyper-exponential distributions, condition (3.8) is *always* true (that is, for any τ), for hypo-exponential distributions including the Erlang distribution (3.8) is *never* true, and for the exponential distribution (3.8) becomes an equality, implying restarts do not help, but also do not hurt.

Let us investigate some probability distributions more formally and evaluate (3.8) for each of them.

3.2.1.1 Exponentially Distributed Task Completion Time

For exponentially distributed task completion time both sides of the inequality are computed as follows.

$$\mathrm{E}[T] = \int_0^\infty t\lambda \cdot e^{-\lambda t}\, dt = \left. -\frac{t}{e^{\lambda t}} - \frac{1}{\lambda}d^{-\lambda t} \right]_0^\infty = \frac{1}{\lambda}$$

$$\mathrm{E}[T - \tau | T > \tau] = \int_\tau^\infty (t - \tau)\lambda \cdot e^{-\lambda(t-\tau)}\, dt =$$

$$= e^{\lambda \tau} \left[-\frac{t}{e^{\lambda t}} - \frac{1}{\lambda}e^{-\lambda t} \right]_\tau^\infty + \left[\tau e^{\lambda(\tau - t)} \right]_\tau^\infty =$$

$$= e^{\lambda \tau} \left(\tau e^{-\lambda \tau} + \frac{1}{\lambda}\tau e^{-\lambda \tau} \right) + \tau = \frac{1}{\lambda}$$

As mentioned above, for the exponential distribution restart is not beneficial, but does not do any harm either. The exponential distribution has a constant hazard rate function $h(t) = \lambda$, which means that task completion is equally likely at all times.

3.2.1.2 Uniform Distributed Task Completion Time

For the uniform distribution on $[a, b]$,

$$f(t) = \begin{cases} 0 & 0 \leq t < a \\ \frac{1}{b-a} & a \leq t \leq b \\ 0 & b < t \end{cases}$$

we obtain

$$E[T] = \frac{a+b}{2}$$

$$E[T - \tau | T > \tau] = \int_\tau^b tf(t - \tau)\, dt = \int_\tau^b t\frac{1}{(b-\tau)}dt - \tau =$$

$$= \frac{1}{(b-\tau)}\frac{1}{2}(b^2 - \tau^2) - \tau = \frac{1}{(b-\tau)}\frac{1}{2}(b+\tau)(b-\tau) - \tau =$$

$$= \frac{(b+\tau)}{2} - \tau = \frac{b}{2} - \frac{\tau}{2}$$

$$= \begin{cases} \frac{1}{2}(a+b) - \tau & \text{for } \tau < a \\ \frac{1}{2}(\tau+b) - \tau = \frac{1}{2}(b-\tau) & \text{for } a \le \tau \le b \\ 0 & \text{for } b < \tau. \end{cases}$$

In all cases $E[T - \tau | T > \tau]$ is smaller than $E[T]$, so one should always wait for the task to complete, which will be faster on the average than a new try after a restart.

3.2.1.3 Weibull Distributed Task Completion Time

The Weibull distribution is very often used in reliability theory because it can model increasing as well as decreasing failure rate. Increasing failure rate shows degradation of system dependability. The density of the Weibull distribution is given by

$$f(t) = \lambda^\alpha \alpha t^{\alpha-1} e^{-(\lambda t)^\alpha} \tag{3.9}$$

with shape parameter α and scale parameter λ (see also Appendix B.2.5). Three cases must be distinguished, $\alpha < 1$, where the hazard rate is decreasing, $\alpha = 1$, where it is constant and $\alpha > 1$, where the hazard rate is increasing. For the expected value and the conditional expectation one obtains

$$E[T] = \frac{1}{\lambda} \cdot \Gamma(1 + \frac{1}{\alpha}) \tag{3.10}$$

$$E[T - \tau | T > \tau] = \frac{\int\limits_\tau^\infty (t-\tau)f(t)\, dt}{\int\limits_\tau^\infty f(t)\, dt} = \tag{3.11}$$

$$= \frac{\int\limits_\tau^\infty tf(t)\, dt - \tau \int\limits_\tau^\infty f(t)\, dt}{\int\limits_\tau^\infty f(t)\, dt} =$$

$$= \frac{\int\limits_\tau^\infty tf(t)\, dt - \tau e^{-(\lambda\tau)^\alpha}}{e^{-(\lambda\tau)^\alpha}} = \frac{1}{\lambda}\Gamma\left(\frac{1+\alpha}{\alpha}, (\lambda\tau)^\alpha\right)e^{(\lambda\tau)^\alpha} - \tau \tag{3.12}$$

The Gamma function $\Gamma(z)$ and the incomplete Gamma function $\Gamma(z, y)$ can be found in Appendix 10. Note that the Gamma function is no probability distribution and does not integrate to one.

A task should be restarted if the inequality $E[T] < E[T - \tau | T > \tau]$ holds. Writing out the condition under which restart should be applied gives

$$E[T] < E[T - \tau | T > \tau] \tag{3.13}$$

$$\Leftrightarrow \quad \Gamma\left(\frac{1+\alpha}{\alpha}\right) < \Gamma\left(\frac{1+\alpha}{\alpha}, (\lambda\tau)^\alpha\right) e^{(\lambda\tau)^\alpha} - \lambda\tau \tag{3.14}$$

$$\Leftrightarrow \quad -\lambda\tau > \int_0^{(\lambda\tau)^\alpha} t^{\frac{1}{\alpha}} e^{-t} \, dt + \int_{(\lambda\tau)^\alpha}^\infty t^{\frac{1}{\alpha}} e^{-t} \, dt \left(1 - e^{(\lambda\tau)^\alpha}\right). \tag{3.15}$$

Analytical derivation of the range of α and λ for which inequality (3.13) holds is not straight forward. Figure 3.3 shows a graphical representation of the solution and the zero plane. If $E[T] - E[T - \tau | T > \tau] < 0$ then restart should be applied while if $E[T] - E[T - \tau | T > \tau] > 0$ waiting for task completion is advised. The value of the scale parameter λ has no effect on the decision. Whether restart is beneficial depends solely on the shape parameter α. For increasing hazard rate ($\alpha > 1$) restart should be applied and for decreasing hazard rate ($\alpha < 1$) no restart should be applied.

If $\alpha = 1$ expression (3.13) is an equation because

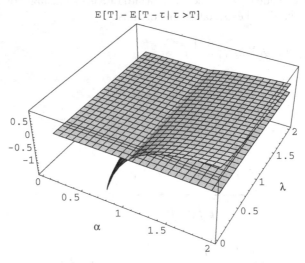

Fig. 3.3 Values greater than zero indicate that restart is beneficial as a function of the shape (α) and scale (λ) parameter of the Weibull distribution

$$E[T] = \frac{1}{\lambda}\Gamma(2) = \frac{1}{\lambda} \qquad (3.16)$$

and

$$E[T - \tau | T > \tau] = \frac{1}{\lambda}\Gamma(2, \lambda\tau)e^{\lambda\tau} - \tau = \frac{1}{\lambda}\int\limits_{\lambda\tau}^{\infty} te^{-t}\, dt \;\; e^{\lambda\tau} - \tau = \quad (3.17)$$

$$= \frac{e^{\lambda\tau}}{\lambda}e^{-\lambda\tau} \cdot (1 + \lambda\tau) - \tau = \frac{1}{\lambda}. \qquad (3.18)$$

This corresponds to the special case of the exponential distribution where restart neither helps nor hurts.

3.2.1.4 Pareto Distributed Task Completion Time

Last, but not least we want to examine the Pareto distribution. The Pareto distributions describes situations with objects, of which most are small, but some are also extremely large, e.g. size of documents, number of links on a page, job size of super computers, word frequencies, response times, etc. The Pareto distribution has density

$$f(t) = \frac{a \cdot b^a}{t^{a+1}}, \qquad \text{for } t > b.$$

For $a \neq 1$, the expected completion time with and without restart evaluate to

$$E[T] = \frac{a \cdot b}{a - 1}$$

$$E[T - \tau | T > \tau] = \int_{\tau}^{\infty} (t - \tau)\frac{b}{(t - \tau)^2}\, dt =$$

$$= [b\ln(t - \tau)]_{\tau}^{\infty} = \infty.$$

Having waited already time τ, the expected completion time does not converge, therefore it is always useful to restart a task with Pareto distributed completion time. Pareto distributed tasks take very long not too frequently, but if they do so, they should be aborted and restarted.

Very common are more complex distributions, such as phase-type distributions [111]. As already mentioned the hyper-exponential distribution always performs better with restarts. It is a distribution of a particular type that seems to be typical for restarts to succeed [132]. These distributions take values from different random variables with different probabilities, that is, with probability p_1 the random variable is distributed as X_1, with probability p_2 distributed as X_2, etc. It then is useful to perform a restart when it gets more likely one drew one of the slower distributions, since then a restart provides a chance to draw one of the faster distributions instead.

3.2.2 When Does Restart Improve the Probability of Meeting a Deadline?

Let the random variable T denote the completion time of a job, with probability distribution $F(t), t \in [0, \infty)$, and probability density function $f(t), t \in [0, \infty)$, as above and let d denote the deadline we set out to meet. We are interested in the probability $F_\tau(d)$ that the deadline is met.

One can intuitively reason about the completion time distribution with restarts as Bernoulli trials. At each interval between restarts there is a probability $F(\tau)$ that the completion 'succeeds.' Hence, if the deadline d is a multiple of the restart time τ, we can relate the probability of missing the deadline without and with restart through:

$$1 - F_\tau(d) = (1 - F(\tau))^{\frac{d}{\tau}}. \qquad (3.19)$$

Then the probability of meeting the deadline under restart is

$$F_\tau(d) = 1 - (1 - F(\tau))^{\frac{d}{\tau}} \qquad (3.20)$$

and restart only makes sense if the probability of meeting the deadline is higher with restart than without:

$$F_\tau(d) > F(d). \qquad (3.21)$$

We can reason differently and say that we have a given time interval of length d which we may split into a number of subintervals. For some distributions the probability of completion within the given time is higher when waiting out the given time whereas for others many tries and waiting shortly is more promising. Following that reasoning equation (3.21) can be reformulated using $d = k\tau$ with $k \in \mathbb{N}$ as

$$1 - (1 - F(\tau))^k > F(k\tau). \qquad (3.22)$$

We will verify or falsify the above condition for some distributions that are commonly used in modelling the lifetime of technical components. All probability distributions are defined in the appendix.

3.2.2.1 Exponentially Distributed Task Completion Time

For exponentially distributed task completion time condition (3.21) is easily proven to become an equality.

$$F_\tau(d) \overset{?}{>} F(d)$$
$$\Leftrightarrow \quad 1 - F(d) \overset{?}{>} (1 - F(\tau))^{\frac{d}{\tau}}$$
$$\Leftrightarrow \quad e^{-\lambda d} \overset{?}{>} e^{-\lambda \tau \cdot \frac{d}{\tau}}$$
$$\Leftrightarrow \quad F_\tau(d) = F(d)$$

The probability of meeting the deadline is equal whether or not one restarts an exponentially distributed job. Hence, for exponential job completion time restart neither helps nor hurts.

3.2.2.2 Hypo-Exponentially Distributed Task Completion Time

For the hypo-exponential distributions (like the Erlang distribution) we can show that condition (3.21) never holds. Let us relate the restart interval τ and the deadline d such that τ is some fraction $\tau = \alpha d$ of the deadline where $0 < \alpha < 1$ is known. In fact, here an Erlang distribution is used for illustration purpose.

$$1 - F(d) \overset{?}{>} (1 - F(\tau))^{\frac{d}{\tau}}$$
$$\Leftrightarrow \quad (1 - F(d))^{\tau} \overset{?}{>} (1 - F(\tau))^{d}$$
$$\Leftrightarrow \quad (1 - F(d))^{\alpha d} \overset{?}{>} (1 - F(\alpha d))^{d}$$
$$\Leftrightarrow \quad (1 - F(d))^{\alpha} \overset{?}{>} (1 - F(\alpha d))$$
$$\Leftrightarrow \quad (1 - 1 + e^{-\lambda d}(1 + \lambda d))^{\alpha} \overset{?}{>} (1 - 1 + e^{-\lambda d}(1 + \alpha \lambda d))$$
$$\Leftrightarrow \quad (1 + \lambda d)^{\alpha} \overset{?}{>} 1 + \alpha \lambda d$$

Assuming that $\lambda d > 0$ (neither a deadline nor an event rate less or equal to zero makes much sense), and for $0 < \alpha < 1$ it has been proven [107] that

$$(1 + \lambda d)^{\alpha} < 1 + \alpha \lambda d \tag{3.23}$$

which means that restart never is beneficial for a completion time distribution of the hypo-exponential type, no matter what deadline is set or what the parameters of the distribution are. This can be interpreted as being a consequence of the more deterministic nature of the hypo-exponential distribution, which has little variance and therefore does not benefit from restart.

3.2.2.3 Weibull Distributed Task Completion Time

When using the Weibull distribution the decision whether restart increases the probability that the task will complete before the deadline or not depends on the value of the shape parameter α as shown in the following.

For the Weibull distribution condition (3.22) is equivalent to

$$1 - \left(1 - e^{-(\lambda k \tau)^\alpha}\right) > \left(1 - \left(1 - e^{-(\lambda \tau)^\alpha}\right)\right)^k$$

$$\Leftrightarrow \quad e^{-(\lambda k \tau)^\alpha} > \left(e^{-(\lambda \tau)^\alpha}\right)^k$$

$$\Leftrightarrow \quad e^{-(\lambda k \tau)^\alpha} > \left(e^{-k\,(\lambda \tau)^\alpha}\right) \tag{3.24}$$

$$\Leftrightarrow \quad k^\alpha < k. \tag{3.25}$$

Inequality (3.25) holds for all values of k iff $\alpha < 1$. If $\alpha = 1$ the Weibull distribution reduces to the exponential distribution and restart neither helps nor does it hurt. The hazard rate of the Weibull distribution is increasing for $\alpha > 1$, for $\alpha = 1$ it is constant and for $\alpha < 1$ it decreases. The shape of the hazard rate function is an indication for whether restart is useful or not.

Increasing hazard rate can be interpreted as an increase in the *potential* of completion, while a decreasing hazard rate indicates a decrease in potential of completion. As the potential of completion decreases it is helpful to restart the task and benefit again of the still higher potential of completion of a short-running task. For increasing hazard rate the reasoning is the equivalent. The longer a task has been running already the higher its potential of completion. A restart would mean to discard the current relatively high potential of completion and instead return to a newly started task with low potential of completion, which is not advisable.

The hazard rate of the Weibull distribution is shown in Appendix B.2.5.

3.3 Conclusions

In the previous subsections we have investigated two different metrics and a number of different distributions. We observed that restart is beneficial for certain distributions and harmful for others, irrespective of the considered metric. Moreover, it can be observed and has been proven in [57] that the shape of a distribution's hazard rate determines whether or not a distribution is amenable to restart. A decreasing hazard rate indicates that restart will improve all metrics. In cases where the hazard rate function is not strictly increasing or decreasing, like the hyper-/hypo-exponential distribution, more detailed analysis is necessary.

Chapter 4
Moments of Completion Time Under Restart

The restart method is based purely on the task completion time. In the previous chapter conditions under which the method is triggered have been investigated. When applying the restart method the only relevant question from a quantitative analysis point is when to restart. For the task under consideration its processing time is monitored and if the processing time exceeds a given value τ then the task is aborted and restarted from beginning. In stochastic terms one may say restart uses a completion time distribution, and a job is restarted when the remaining expected completion time is longer than the expected completion time when restarting the job, taking into account the time already waited for completion. This is similar to age replacement in preventive maintenance.

There are two important issues which we have not yet considered and which will be addressed in this chapter for the metric moments of completion time. Those are: (1) to quantify how much the moments of completion time can be improved by restarting the job and (2) when to restart as to achieve the most improvement. These issues are also addressed in [163, 168, 165, 167].

In order to give expressions answering (1) we need to formulate mathematical models for the moments of completion time under restart. Then (2) is provided by optimising the moments of completion time under restart with respect to the restart interval length.

We have to distinguish whether a finite or an infinite number of restarts is allowed and we treat the first moment separately as it turns out to be a much simpler special case, allowing for much simpler optimisation algorithms.

But before entering into the technical details of restart we investigate in the next subsection whether optimising, that is in our case *minimising*, the moments of completion time improves system operation. We study the meaning of the moments for the shape of a probability distribution and how reducing the moments changes the shape of a distribution and hence the potential outcome of a random variable drawn from the modified distribution.

K. Wolter, *Stochastic Models for Fault Tolerance*,
DOI 10.1007/978-3-642-11257-7_4, © Springer-Verlag Berlin Heidelberg 2010

4.1 The Information Captured by the Moments of a Distribution

The models and methods provided can be used to compute and optimise all moments of completion time under restart. However, we would like to be able to interpret the moments when we want to see how the shape of a distribution, and in particular the mass distribution, is changed when changing the distribution's moments. Unfortunately, the *raw moments*, which we optimise, do not so easily translate into characteristics such as variance, skewness and kurtosis, that are based on the *central moments* of a distribution. Geometric interpretations are in any case only known for combinations of the first four central moments and we will only discuss those here.

Let T be a random variable describing some job execution time with probability distribution $f(t)$. Then its n−th raw moment can be expressed as the expected value $\mathrm{E}\left[T^n\right]$ with

$$\mu_n = \mathrm{E}\left[T^n\right] = \int_0^\infty t^n \, f(t) \, dt.$$

The n−th central moment μ'_n is defined as

$$\mu'_n = \int_0^\infty (t - \mu_1)^n \, f(t) \, dt.$$

The raw moments and the central moments can each be expressed in one another [1]. We describe the central moments in the raw moments as

$$\mu'_n = \sum_{k=0}^n \binom{n}{k}(-1)^{n-k}\mu_k\mu_1^{n-k}. \tag{4.1}$$

Using $\mu_0 = 1$ the first five central moments evaluate to

$$\mu'_1 = 0$$
$$\mu'_2 = -\mu_1^2 + \mu_2$$
$$\mu'_3 = 2\mu_1^3 - 3\mu_1\mu_2 + \mu_3$$
$$\mu'_4 = -3\mu_1^4 + 6\mu_1^2\mu_2 - 4\mu_1\mu_3 + \mu_4$$
$$\mu'_5 = 4\mu_1^5 - 10\mu_1^3\mu_2 + 10\mu_1^2\mu_3 - 5\mu_1\mu_4 + \mu_5$$

The first moment of a probability distribution is its expected value $\mathrm{E}[T]$, estimated by the mean of a sample. Obviously, the smaller the expected value the smaller are in many cases the observations. If we can reduce the expected job execution time we will in most cases in fact see shorter job execution times. Minimising the first moment therefore is certainly worth while.

The second moment of a probability distribution is a measure describing the degree of variation. The variance, the second central moment above, is defined through first and second raw moment and it is not obvious whether minimising those will also reduce the variance. All we can say, as we will see later, is that if restart at τ_1 minimises the first raw moment and restart at τ_2 minimises the second raw moment, then the variance with restarts at time τ_2 is less than the variance with restarts at time τ_1.

The probability distributions we consider typically have strictly non-negative support, meaning they are bounded from below by zero and high variation is rather unpleasant since it means that very large values are still reasonably likely to occur. Decreasing variance will typically mean that large extremes are less likely. This again will in some cases reduce job execution times.

A metric based on the third central moment is the skewness v, defined as

$$v = \frac{\mu'_3}{\mu'^{3/2}_2}$$

$$= \frac{2\mu_1^3 - 3\mu_1\mu_2 + \mu_3}{(-\mu_1^2 + \mu_2)^{3/2}}$$

The skewness v of a distribution indicates the asymmetry of the distribution around its mean, characterising the shape of the distribution [91]. Symmetric distributions like the normal distribution have $v = 0$, whereas for the exponential distribution, e.g., $v = 2$. The skewness of the Gamma distribution is

$$v_{\text{Gamma}} = \frac{2}{\alpha}.$$

Special cases are obtained for certain values of α. The exponential distribution is a Gamma distribution with $\alpha = 1$, and the Erlang distribution is a Gamma distribution with $\alpha \geq 1$, $\alpha \in \mathbb{N}$, where α are the number of phases of the Erlang distribution. We see that the skewness of the Erlang distribution decreases with increasing number of phases and we know that the Erlang distribution becomes *more deterministic* with increasing number of phases.

A negative skewness means that the distribution is skewed to the left, i.e. has the long flat tail to the left.

The kurtosis β_2 and *kurtosis excess* γ_2 [1] of a distribution are defined as a normalised form of the fourth central moment which describes the degree of peakedness of a distribution. γ_2 is scaled such that the normal distribution has kurtosis excess zero. Different definitions of kurtosis exist. We use

$$\beta_2 = \frac{\mu'_4}{\mu'^2_2}$$

and $\gamma_2 = \beta_2 - 3$. In terms of the raw moments it is

$$\beta_2 = \frac{-3\mu_1^4 + 6\mu_1^2\mu_2 - 4\mu_1\mu_3 + \mu_4}{(-\mu_1^2 + \mu_2)^2} \tag{4.2}$$

The Gamma distribution has $\gamma_2 = \frac{6}{\alpha}$, with again the exponential distribution being the special case $\alpha = 1$ and the Erlang distribution having $\alpha \geq 1, \alpha \in \mathbb{N}$.

It is not obvious how minimising the raw moments affects the kurtosis. One can assume that minimising the first moment will maximise the kurtosis, since the first moment appears in the highest power in (4.2) and hence has the greatest impact. But we may reason that for our purpose of reducing completion times a deterministic distribution is desirable and hence reducing the kurtosis of the completion time distribution is beneficial.

4.2 Models for Moments of Completion Time

In this section we define the stochastic models for completion time under restart. For an unbounded number of allowed restarts we are able to derive an elegant formula for all moments, which we can then use to compute all moments iteratively. The moments can be bounded from below and above by using a simple geometric approximation.

We do not address choosing an appropriate restart interval length, but assume that restarts happen in fixed-length intervals. We will see later that optimal higher moments of completion time are often rather achieved when using restart intervals of individual and different length. We show the degree of improvement depending on the restart interval length where all intervals are equal.

An expression for the moments of completion time can also be derived if the number of allowed restarts is finite. Based on this expression an algorithm is formulated that goes backward in time to compute all moments for the case of a finite number of restarts.

Optimisation of the moments of completion time with respect to the restart interval length is the topic of the next section.

4.2.1 Unbounded Number of Restarts

Let the random variable T represent the completion time of a job without restarts, $f(t)$ its probability density function, and $F(t)$ its distribution. For convenience, but without loss of generality,[1] we assume that $F(t)$ is a continuous probability distribution function defined over the domain $[0, \infty)$, so that $F(t) > 0$ if $t > 0$. Assume τ is a restart time,[2] and the overhead associated with restarting is c time

---

[1] At the cost of more notation, and with a proper discussion for special cases, the results in this section also apply to distributions defined over finite domains, as well as to defective distributions and distributions with jumps.

[2] At times we also refer to τ as the as restart *interval*.

units for each restart (we also refer to c as the 'cost' of a restart). We introduce the random variable T_τ to denote the completion time when an *unbounded* number of restarts is allowed. That is, a restart takes place periodically, every τ time units, until completion of the job. We write $f_\tau(t)$ and $F_\tau(t)$ for the density and distribution of T_τ, and we are interested in the moments of T_τ, and later also in the optimal value of the restart time τ itself. To formally derive an expression for the moments, we first need an expression for the distribution and density of the completion time with restarts. We assume that a restart preempts the previous attempt, and that the completion times for consecutive attempts are statistically identical and independent. One can then reason about completion of a task in a restart interval as a Bernoulli trial with success probability $F(\tau)$. That is, the completion time with restarts relates to that without restarts as:

$$F_\tau(t) = \begin{cases} 1 - (1 - F(\tau))^k (1 - F(t - k(\tau + c))) & \text{if } k(\tau + c) \le t < k(\tau + c) + \tau \\ 1 - (1 - F(\tau))^{k+1} & \text{if } k(\tau + c) + \tau \le t < (k + 1)(\tau + c) \end{cases}$$
$$(4.3)$$

for $k = 0, 1, 2, \ldots$. For the density we obtain for any integer value $k = 0, 1, 2, \ldots$:

$$f_\tau(t) = \begin{cases} (1 - F(\tau))^k f(t - k(\tau + c)) & \text{if } k(\tau + c) \le t < k(\tau + c) + \tau \\ 0 & \text{if } k(\tau + c) + \tau \le t < (k + 1)(\tau + c) \end{cases}$$
$$(4.4)$$

As an example we will use the hyper/hypo-exponential distribution (see Appendix B for the precise mathematical characterisation), as it represents typical almost bimodal behaviour. The random variable draws with probability $p = 0.9$ from an Erlang distribution with two phases and mean 0.1 and with probability 0.1 from an Erlang distribution with two phases and mean 1.0.

Figure 4.1 shows the density of the mixed hyper/hypo-exponential distribution. It turns out that for a single restart, the optimal restart time is about 0.25, while for unbounded number of repeated restarts, the optimal restart time is about 0.19. Both values are indeed not too far above the mean 0.1 of the first Erlang distribution. The

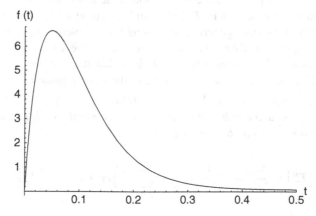

Fig. 4.1 The probability density function of the mixed hyper/hypo-exponential distribution. Optimal restart times are 0.25 for single restart and 0.19 for unbounded number of restarts

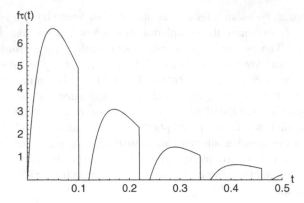

Fig. 4.2 The probability density f_τ of completion time with unbounded number of restarts (based on hyper/hypo-exponentially distributed completion time without restarts, with restart time $\tau = 0.1$ and cost $c = 0.02$)

expected completion time decreases for a single restart from 0.190 to 0.136, and for unbounded restarts to 0.127, see also Fig. 4.6 on p. 64.

It is worth visualising the density of T_τ, see Fig. 4.2 for a mixed hyper/hypo-exponentially distributed T, with parameters as in Appendix B.2.4, restart time $\tau = 0.1$, and cost $c = 0.02$. In what follows, we also need the partial moments $M_n(\tau)$ at τ of the completion time, which are defined as:

$$M_n(\tau) = \int_0^\tau t^n f(t)dt = \int_0^\tau t^n f_\tau(t)dt. \tag{4.5}$$

The equality of partial moments of T and T_τ follows from the fact that their respective densities are identical between 0 and τ (see (4.4) for $k = 0$).

We will exploit the structure of (4.4) to obtain computationally attractive expressions for the moments of T_τ, and to gain further insight into optimal restart policies. We expected to find in the literature or text books the following rather straightforward derivation of the moments $E\left[T_\tau^n\right]$ of completion time T_τ, but did not succeed. The literature provides more generic expressions for completion time (e.g., [113]), but these are computationally not very attractive. Hence, we give the derivation and recursive expressions here, although it is obvious that the authors of for instance [104] and [132] must have worked from some of the same equations.

Theorem 4.1 *The moments* $E\left[T_\tau^n\right] = \int_0^\infty t^n f_\tau(t)dt$, $n = 1, 2, \ldots$, *of the completion time with unbounded number of restarts, restart interval length* $\tau > 0$, *and time* c *consumed by a restart, can be expressed as:*

$$E\left[T_\tau^n\right] = \frac{M_n(\tau)}{F(\tau)} + \frac{1 - F(\tau)}{F(\tau)} \sum_{l=0}^{n-1} \binom{n}{l}(\tau + c)^{n-l} E\left[T_\tau^l\right], \tag{4.6}$$

where $E\left[T_\tau^0\right] = 1$.

Proof The derivation is particularly elegant if one exploits the recursive structure of
(4.4). First, by definition, we have:

$$E\left[T_\tau^n\right] = \int_0^\infty t^n f_\tau(t)dt$$

$$= \int_0^\tau t^n f_\tau(t)dt + \int_{\tau+c}^\infty t^n f_\tau(t)dt = M_n(\tau) + \int_{\tau+c}^\infty t^n f_\tau(t)dt. \quad (4.7)$$

Then, we use that from (4.4) it follows that for any $t \geq 0$,

$$f_\tau(t + \tau + c) = (1 - F(\tau))f_\tau(t),$$

and thus:

$$\int_{\tau+c}^\infty t^n f_\tau(t)dt = \int_0^\infty (t+\tau+c)^n f_\tau(t+\tau+c)dt = (1-F(\tau))\int_0^\infty (t+\tau+c)^n f_\tau(t)dt. \quad (4.8)$$

Combining (4.7) and (4.8) we obtain:

$$E\left[T_\tau^n\right] = M_n(\tau) + (1 - F(\tau))\int_0^\infty (t + \tau + c)^n f_\tau(t)dt,$$

which we write out as:

$$E\left[T_\tau^n\right] = M_n(\tau) + (1 - F(\tau))\int_0^\infty \sum_{l=0}^n \binom{n}{l}(\tau + c)^{n-l}t^l f_\tau(t)dt =$$

$$M_n(\tau) + (1 - F(\tau))\sum_{l=0}^n \binom{n}{l}(\tau + c)^{n-l}E\left[T_\tau^l\right].$$

One then solves this equation for $E\left[T_\tau^n\right]$, cancelling out the highest moment within
the sum, to obtain:

$$E\left[T_\tau^n\right] = \frac{M_n(\tau)}{F(\tau)} + \frac{1 - F(\tau)}{F(\tau)}\sum_{l=0}^{n-1} \binom{n}{l}(\tau + c)^{n-l}E\left[T_\tau^l\right].$$

□

For example, the expected completion time is given by:

$$E\left[T_\tau\right] = \frac{M_1(\tau)}{F(\tau)} + \frac{1 - F(\tau)}{F(\tau)}(\tau + c). \quad (4.9)$$

The expression for the variance can also be found in [104]. The result for the first
moment is indeed as it should be: (4.9) must account for the interval in which the

task completes, as well as for the times the job fails to complete. The first term in
(4.9) is the expected download time conditioned on success within a restart interval.
The second term equals the interval length $\tau + c$ times the expected value of a
modified geometric distribution [69] with parameter $F(\tau)$, since, indeed, in every
interval the probability of completion is $F(\tau)$.

Finally, note that by requiring that $\tau > 0$ in Theorem 4.1, the denominator $F(\tau)$
in (4.6) is positive, since we assumed continuous distributions defined over $[0, \infty)$.
For $\tau \downarrow 0$ and $c = 0$, we can apply l'Hospital's rule, to see that $\mathrm{E}[T_\tau] \to f^{-1}(0)$,
which tends to infinity for our running example meaning that if we do an unbounded
number of immediate restarts the job will never complete.

Equation (4.6) directly yields an algorithm to iteratively compute all moments up
to some specified value N. We reformat it here as an algorithm for completeness of
the presentation:

Algorithm 1 (Computation of all Moments, Unbounded Restarts)

> Set $\mathrm{E}\left[T_\tau^0\right] = 1$ for chosen $\tau > 0$;
> For $n = 1$ to N {
>   compute $\mathrm{E}\left[T_\tau^n\right] = \frac{M_n(\tau)}{F(\tau)} + \frac{1-F(\tau)}{F(\tau)} \sum_{l=0}^{n-1} \binom{n}{l}(\tau + c)^{n-l}\mathrm{E}\left[T_\tau^l\right]$
> }

Using this basic algorithm we obtained the results as shown in Fig. 4.3 through 4.13.
Figure 4.3 provides the relative gain depending on the restart interval length using
restarts for the first three moments, where we use identical interval lengths for all
restarts, which we will later see is not always optimal. Notice that the gain increases
rather dramatically with the order of the moment. Also, notice the wide range of
restart times which perform well, which suggests that rough estimates may often
suffice to set a restart time. An important engineering rule is not to take the restart

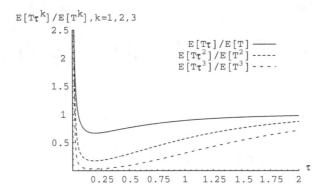

Fig. 4.3 Restart time versus the normalised difference between unbounded restarts and no restarts,
for first three moments

time too small, since for many realistic distributions, the completion time will then tend to infinity (the hyper-exponential distribution being one exception). In general, it may be safer to take the restart time too large than too small.

As a corollary of Theorem 4.1 we state a fundamental result, which was also observed by [27] for failure detectors.

Corollary 4.2 *Under unbounded restarts, the expectation (as well as higher moments) of the completion time T_τ with restart time $\tau > 0$ (for which $F(\tau) > 0$), is always finite, even if the moments of the original completion time are not.*

This is an important observation, stressing the value of restarts for situations in which there is a (strictly) positive probability that a task can fail (thus making the moments of completion time infinite).

Geometric approximation. The results obtained above also suggest bounds for the moments by using the (modified) geometric distribution (see also expression (4.2.12) in [57] for the dual result in terms of mean time between failures). To bound the first moment, one replaces the first term in (4.6), which refers to the interval in which the job completes, by its upper and lower bounds 0 and τ, respectively. This can be generalised to all moments, using two discrete random variables A_τ and B_τ, with

$$A_\tau = k(\tau + c), \text{with probability } (1 - F(\tau))^k F(\tau), \quad k = 0, 1, \ldots,$$
$$B_\tau = k(\tau + c) + \tau, \text{with probability } (1 - F(\tau))^k F(\tau), \quad k = 0, 1, \ldots. \quad (4.10)$$

Since we know from (4.3) that

$$k(\tau + c) \le T_\tau \le k(\tau + c) + \tau, \text{with probability } (1 - F(\tau))^k F(\tau), \quad k = 0, 1, \ldots,$$

we have that $\mathrm{E}\left[A_\tau^n\right] \le \mathrm{E}\left[T_\tau^n\right] \le \mathrm{E}\left[B_\tau^n\right]$, for $n = 1, 2, \ldots$. Note that A_τ has a modified geometric distribution [69] and that $B_\tau = A_\tau + \tau$. Figure 4.4 shows $\mathrm{E}[T_\tau]$ as well as the bounds for the mixed hyper/hypo-exponential distribution. The bounds, whose summed error equals exactly τ, are excellent approximations as long

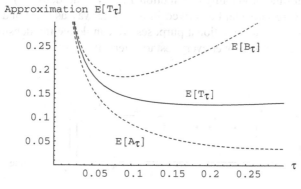

Fig. 4.4 Approximation of expected completion time using geometric distributions

as the restart time τ is small relative to the mean completion time $E[T_\tau]$. For the area around the optimal restart time the bounds are not particularly tight. Nevertheless, the geometric approximation may prove very useful for determining a conservative restart time. For instance, for Fig. 4.4, the optimal restart time (that is, the minimum of the curve) for the upper bound is 0.09, and for the lower bound 0.31, while the real optimum lies in between (namely at 0.20). Moreover, Fig. 4.3 shows that for $\tau = 0.31$, the expected completion time is still close to optimal. Using 0.31 as a conservative restart time is also consistent with the above-mentioned engineering rule that it is better to restart too late than too early.

4.2.2 Finite Number of Restarts

There may be cases in which one is interested in a *finite* number of restarts. For example, in the mixed hyper/hypo-exponential example, too low a restart time is very detrimental for the completion time if there is no bound on the number of restarts. Although this need not generally be the case (the hyper-exponential distribution is a counter example), for many distributions it may be wise to limit the number of restarts, or increase the period between restarts with the restart count. This leads to a situation with finite and non-identical restart intervals, for which we derive an algorithm to compute all moments. Perhaps one would expect that restarts should take place with fixed-length intervals between them, but we will see that this is often not optimal. We provide an algorithm to compute the optimal restart times for the first moment. We explain why the first moment is considerably simpler to optimise than higher moments, for which we do not know an algorithm with proven convergence.

For our discussion it is convenient to label the restarts as shown in Fig. 4.5. We assume the total number of restarts is K, and the restart intervals have length $\tau_1, \tau_2, \ldots, \tau_{K-1}, \tau_K$, respectively. The k-th interval starts at time s_k. So, we get $s_1 = 0$, $s_2 = \tau_1 + c$, $s_3 = \tau_1 + c + \tau_2 + c$, etc., until $s_K = \sum_{k=1}^{K} \tau_k + Kc$. The completion time with K restarts is represented by the random variable $T_{\tau_1,\ldots,\tau_K}$. The completion time probability distribution $F_{\tau_1,\ldots,\tau_K}$ and density $f_{\tau_1,\ldots,\tau_K}$ for the scenario with K restarts can be derived in the same way as (4.3) and (4.4). If we introduce $\tau_{K+1} = \infty$ for notational purposes we can define the density and distribution function piece-wise over every restart interval:

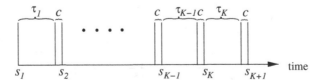

Fig. 4.5 Labeling restarts, total of K restarts

$$F_{\tau_1,\ldots,\tau_K}(t) = \begin{cases} 1 - \prod_{i=1}^{k-1}(1 - F(\tau_i))(1 - F(t - s_k)) & \text{if } s_k \le t < s_k + \tau, k = 1, \ldots, K+1 \\ 1 - \prod_{i=1}^{k}(1 - F(\tau_i)) & \text{if } s_k + \tau_k \le t < s_{k+1}, k = 1, 2, \ldots, K \end{cases}$$

$$f_{\tau_1,\ldots,\tau_K}(t) = \begin{cases} \prod_{i=1}^{k-1}(1 - F(\tau_i))f(t - s_k) & \text{if } s_k \le t < s_k + \tau, k = 1, \ldots, K+1 \\ 0 & \text{if } s_k + \tau_k \le t < s_{k-1}, k = 1, 2, \ldots, K \end{cases} \tag{4.11}$$

As for the unbounded case, we express the moments in the following theorem in a manner convenient for computational purposes. This time, we express the moments of the completion time with K restarts in that with one restart less.

Theorem 4.3 *The moments $E\left[T_{\tau_1,\ldots,\tau_K}^n\right] = \int_0^\infty t^n f_{\tau_K,\ldots,\tau_1}(t)dt$, $n = 1, 2, \ldots$, of the completion time with K restarts, restart interval lengths $\tau_1, \tau_2, \ldots, \tau_K$, and time c consumed by each restart, can be expressed as:*

$$E\left[T_{\tau_1,\ldots,\tau_K}^n\right] = M_n(\tau_1) + (1 - F(\tau_1)) \sum_{l=0}^{n} \binom{n}{l}(\tau_1 + c)^{n-l} E\left[T_{\tau_2,\ldots,\tau_K}^l\right], \tag{4.12}$$

where $E\left[T_{\tau_2,\ldots,\tau_K}^0\right] = 1$.

Proof The derivation is similar to that of Theorem 4.1. Start from the fact that from (4.11) it follows that for $t \ge 0$:

$$f_{\tau_1,\ldots,\tau_K}(s_2 + t) = (1 - F(\tau_1))f_{\tau_2,\ldots,\tau_K}(t),$$

and then follow the same derivation as in Theorem 4.1. Only the last step, in which $E\left[T_\tau^n\right]$ is solved, has no counterpart in the current proof. □

As an illustration, we get for the first moment:

$$E\left[T_{\tau_1,\ldots,\tau_K}\right] = M_1(\tau_1) + (1 - F(\tau_1))\left(\tau_1 + c + E\left[T_{\tau_2,\ldots,\tau_K}\right]\right). \tag{4.13}$$

The above theorem implies that if $\tau_1, \ldots, \tau_K$ are known beforehand, one can iteratively compute $E\left[T_{\tau_1,\ldots,\tau_K}^N\right]$ for any $N > 0$ by going backward in time. That is, starting from the moments $E\left[T_{\tau_K}^n\right]$, $n = 1, \ldots, N$, one obtains $E\left[T_{\tau_{K-1},\tau_K}^n\right]$, until $E\left[T_{\tau_1,\ldots,\tau_K}^N\right]$. The algorithm thus goes as follows:

Algorithm 2 (Backward Algorithm, first N Moments, K Restarts)

```
For  n = 0  to  N
    Set  E[T^n_{τ_{K+1},...,τ_K}] = E[T^n];
For  k = K  to  1 {
    For  n = 0  to  N {
        E[T^n_{τ_k,...,τ_K}] = M_n(τ_k) + (1 − F(τ_k)) Σ^n_{l=0} (n choose l)(τ_k + c)^{n−l} E[T^l_{τ_{k+1},...,τ_K}];
    }
}
```

A nice feature of the backward algorithm is that it computes moments of subsets $\{\tau_k, \ldots, \tau_K\}$ along the way. One should be careful to interpret those correctly: the moments $\mathrm{E}\left[T^l_{\tau_k,\ldots,\tau_K}\right]$ are for $s_k = 0$, that is, for the case that completion time starts counting at the k-th interval, not before. This feature of the backward algorithm turns out to be its pitfall as well, if we try to use the algorithm for optimisation purposes. The issue is that for higher moments optimisation of the k-th restart time depends on *all* other restart times. Only for the first moment, the optimal value of the k-th restart is insensitive to what happens *before* the k-th restart (that is, to the restarts we labelled $1, \ldots, k - 1$). As a consequence, for the first moment, we can optimise the restart intervals concurrently with computing moments using the backward algorithm.

4.3 Optimal Restart Times for the Moments of Completion Time

In this section we find optimal restart times, which are the times when to restart as to optimise the moments of completion time. We have to distinguish the first moment of completion time and higher moments, as well as an infinite number of allowed restarts and a finite number of restarts. We will see that for all four cases the algorithms for obtaining the optimal restart times are different as are the best restart times themselves.

4.3.1 Expected Completion Time

At first we study the first moment of completion time under restart and develop algorithms that help finding the restart times that will optimise the first moment of completion time. Allowing an unbounded number of restarts we find an interesting relationship between the hazard rate of the completion time and the expected completion time (Theorem 4.4): the inverse hazard rate evaluated at the optimal restart time τ equals the expected completion time under restart, if the cost associated with restart is zero.

We are able to formulate a condition on completion time distributions to be amenable to restart, and a monotonicity relation for the mean completion time as function of the number of restarts (see Sect. 4.3.1.3).

4.3.1.1 Unbounded Number of Allowed Restarts

The optimal restart interval size for the first moment and an unbounded number of allowed restarts can be computed in one optimisation during the computation of the moments under restart with the backward algorithm 2 on the preceding page. It is obtained in a straight forward manner by minimising (4.9) with respect to τ. All optimal restart intervals have equal length.

We give here an implicit relation for the optimal restart time τ^* for the first moment of T_τ. This implicit expression provides us with interesting insight into how the hazard rate of a distribution determines the optimal completion time under restarts. It will also help us to refute the claims on the existence of a cusp point in [104], as we will show in Sect. 4.3.2.

Theorem 4.4 *The optimal restart time $\tau^* > 0$ that minimises the expected completion time $E[T_\tau]$ is such that:*

$$\frac{1 - F(\tau^*)}{f(\tau^*)} = E[T_{\tau^*}] + c. \tag{4.14}$$

That is, if $c = 0$, the inverse of the hazard rate at τ^* equals the expected completion time under unbounded restarts.

Proof To obtain this result, we equate to zero the derivative with respect to τ of $E[T_\tau] = \frac{M_1(\tau)}{F(\tau)} + \frac{1-F(\tau)}{F(\tau)}(\tau + c)$ (the base relation (4.9)):

$$\frac{d}{d\tau}E[T_\tau] = 0 \iff \frac{\tau f(\tau)F(\tau) - f(\tau)M_1(\tau)}{F^2(\tau)} + \frac{1 - F(\tau)}{F(\tau)} - \frac{f(\tau)(\tau + c)}{F^2(\tau)} = 0$$

$$\iff \frac{1 - F(\tau)}{F(\tau)} = \frac{f(\tau)}{F(\tau)}\left(\frac{(\tau + c)}{F(\tau)} - \tau + \frac{M_1(\tau)}{F(\tau)}\right)$$

$$\iff \frac{1 - F(\tau)}{f(\tau)} = \frac{1 - F(\tau)}{F(\tau)}(\tau + c) + c + \frac{M_1(\tau)}{F(\tau)}.$$

Applying (4.9) again we obtain:

$$\frac{d}{d\tau}E[T_\tau] = 0 \iff \frac{1 - F(\tau)}{f(\tau)} = E[T_\tau] + c.$$

$\square$

It is important to realise that (4.14) may hold for many restart values, including $\tau \to \infty$, and not only holds for the global optimum, but also for local minima and maxima. For instance, in Fig. 4.6, the inverse hazard rate indeed crosses $E[T_\tau]$ in

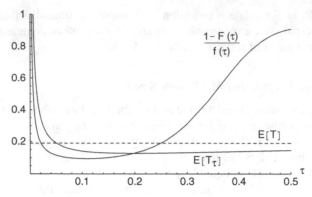

Fig. 4.6 Extrema for the mean completion time are found at restart times τ for which the inverse hazard rate equals $\mathrm{E}[T_\tau]$

its minimum, which gives $\tau^* \approx 0.2$, but also meets at point 0, where the maximum of the completion time under restarts is reached.

As we remarked in Sect. 1.3, there exists a dual problem formulation in terms of maximising the time to failure, and thus, Theorem 4.4 not only holds for restart times that minimise completion time, but also for preventive maintenance times that maximise the mean time to failure (which is meaningful only for certain distributions and without cost (i.e., $c = 0$)).

4.3.1.2 Finite Number of Allowed Restarts

For the first moment, we can optimise the restart intervals concurrently with computing moments using the backward algorithm. We extend the backward algorithm for optimisation of a finite number of restart intervals to the *backward optimisation algorithm*.

The backward optimisation algorithm only requires a single run of K steps; it works backward in time and finds optimal restart times $\tau_K^*, \ldots, \tau_1^*$, in that order.

Algorithm 3 (Backward Optimisation Algorithm, First Moment, K Restarts)

```
Set E[T_{τ*_{K+1},...,τ*_K}] = E[T];
For k = K to 1 {
        compute τ*_k, the value of τ_k > 0 that
        minimises M₁(τ_k) + (1 − F(τ_k))(τ_k + c + E[T_{τ*_{k+1},...,τ*_K}]);
        Set E[T_{τ*_k,...,τ*_K}] = M₁(τ*_k) + (1 − F(τ*_k))(τ*_k + c + E[T_{τ*_{k+1},...,τ*_K}]);
}
```

As already mentioned in Sect. 4.3.1.2 the backward algorithm computes moments of subsets $\{\tau_k, \ldots, \tau_K\}$ on the way. But the moments $E\left[T^l_{\tau_k,\ldots,\tau_K}\right]$ are for $s_k = 0$, that is for s_k shifted on the time axis to zero. The first moment is insensitive to those shifts, i.e. it is insensitive to what happens before the k−th restart. This does not hold true for higher moments. The insensitivity of the first moment of completion time to shifts is proven in the following theorem.

This theorem is necessary to show that the backward optimisation algorithm finds the restart times that optimise the first moment of completion time.

Correctness of the Backward Optimisation Algorithm The following theorem demonstrates that if the backward optimisation algorithm (Algorithm 3) computes a restart time τ_k^* it is optimal, irrespective of the values of τ_1 to τ_{k-1} computed later in the algorithm.

Theorem 4.5 *Assume a restart strategy with K restarts at times $\tau_1, \ldots, \tau_{K-1}, \tau_K$. For the first moment $E\left[T_{\tau_1,\ldots,\tau_K}\right]$, for any value of k between 1 and K, the restart time τ_k^*, that minimises $E\left[T_{\tau_k,\ldots,\tau_K}\right]$, also minimises $E\left[T_{\tau_1,\ldots,\tau_K}\right]$ (and is thus independent of $\tau_1, \ldots, \tau_{k-1}$). For higher moments this is not generally true.*

Proof $E\left[T_{\tau_1,\ldots,\tau_K}\right]$ and $E\left[T_{\tau_k,\ldots,\tau_K}\right]$ relate as follows:

$$E\left[T_{\tau_1,\ldots,\tau_K}\right] = \int_0^{s_k} tf_{\tau_1,\ldots,\tau_K}(t)dt + \prod_{i=1}^{k-1}(1 - F(\tau_i)) \int_{s_k}^{\infty} tf_{\tau_k,\ldots,\tau_K}(t - s_k)dt =$$

$$\int_0^{s_k} tf_{\tau_1,\ldots,\tau_K}(t)dt + \prod_{i=1}^{k-1}(1 - F(\tau_i))E\left[s_k + T_{\tau_k,\ldots,\tau_K}\right](4.15)$$

Since $E\left[s_k + T_{\tau_k,\ldots,\tau_K}\right] = s_k + E\left[T_{\tau_k,\ldots,\tau_K}\right]$, and since the integral between 0 and s_k is independent of τ_k, we see immediately that the value of s_k is immaterial for the optimal restart time τ_k^*:

$$\frac{d}{d\tau_k}E\left[T_{\tau_1,\ldots,\tau_K}\right] = 0 \iff \frac{d}{d\tau_k}E\left[s_k + T_{\tau_k,\ldots,\tau_K}\right] = 0 \iff \frac{d}{d\tau_k}E\left[T_{\tau_k,\ldots,\tau_K}\right] = 0.$$

Hence, τ_k^* is an extreme for $E\left[T_{\tau_1,\ldots,\tau_K}\right]$ if and only if it is an extreme for $E\left[T_{\tau_k,\ldots,\tau_K}\right]$. Even stronger, by inspection of (4.15) it follows that if τ_k^* is an element of the global minimum $\{\tau_k^*, \ldots, \tau_K^*\}$ for $E\left[T_{\tau_k,\ldots,\tau_K}\right]$ (with k restarts) then no other value of τ_k can result in a better expected overall completion time $E\left[T_{\tau_1,\ldots,\tau_K}\right]$ (with K restarts). Thus, τ_k^* must then also be an element of the global optimum $\{\tau_1^*, \ldots, \tau_K^*\}$ for $E\left[T_{\tau_1,\ldots,\tau_K}\right]$. For higher moments, we can not remove the dependence on s_k, and the backward optimisation algorithm does therefore not extend to higher moments. As an example, $E\left[(s_k + T_{\tau_k,\ldots,\tau_K})^2\right]$ leaves a term with $s_k\frac{d}{d\tau_k}E\left[T_{\tau_1,\ldots,\tau_K}\right]$ in the derivative. Hence, restart times $\tau_k, \ldots, \tau_K$ that optimise $E\left[T^2_{\tau_k,\ldots,\tau_K}\right]$ may not optimise $E\left[(s_k + T_{\tau_k,\ldots,\tau_K})^2\right]$ and/or $E\left[T^2_{\tau_1,\ldots,\tau_K}\right]$.   □

As an illustration, we apply the backward optimisation algorithm to our mixed hyper/hypo-exponential distribution, with parameters as given in Appendix B.2.4,

Table 4.1 Optimal restart intervals for finite and unbounded number of restarts

Interval index k	Optimal length τ_k of k-th interval
Unbounded	0.198254
1	0.198254
2	0.198254
3	0.198254
4	0.198256
5	0.198265
6	0.1983
7	0.199
8	0.200
9	0.209
10	0.249

to obtain the optimal restart times given in Table 4.1. Note that the values in the table imply that, for instance, for $k = 9$, the two remaining restart times will be after 0.209 and $0.209 + 0.249 = 0.458$ time units, respectively. As one sees from Table 4.1, the restart intervals have different lengths, longer if it is closer to the last restart. Furthermore, the more restarts still follow, the closer the interval length is to the optimum for unbounded restarts, which is 0.198254. This is as expected.

4.3.1.3 Characteristics of Probability Distributions and Optimal Restart Policies

The backward algorithm (Algorithm 2 in Sect. 4.3.1.2) provides us with machinery to further characterise necessary and sufficient conditions for a random variable T to benefit from restarts. We will see that for the mean completion time, the intuitive condition we derived in Sect. 3.2.1 for a single restart is necessary and sufficient for any number of restarts to be useful. We will also show that if a single restart improves the mean completion time, multiple restarts perform even better, and unbounded restarts performs best.

We use the random variable $T_{\tau K}$ to denote the completion time under K restarts at times $\tau+c, 2(\tau+c), \ldots, K(\tau+c)$, and for technical reasons also use the notation $T_{\tau 0} = T$, for the case without restarts.

Theorem 4.6 *The mean completion time under zero ($E[T]$), $K \geq 1$ ($E\left[T_{\tau K}\right]$) and unbounded restarts ($E[T_\tau]$) interrelate as follows:*

$$E[T_\tau] < \ldots < E\left[T_{\tau K+1}\right] < E\left[T_{\tau K}\right] < \ldots < E[T] \iff E[T_\tau] < E[T],$$
(4.16)

and

$$E[T + c] < E[T - \tau | T > \tau] \iff E[T_\tau] < E[T].$$
(4.17)

Proof The first result follows from the backward algorithm, which uses (4.13) for the first moment. (Note that $T_{\tau K}$ is in fact identical to $T_{\tau_1,\ldots,\tau_K}$ with $\tau_1 = \ldots = \tau_K = \tau$.) If we introduce

$$C_\tau = M_1(\tau) + (1 - F(\tau))(\tau + c),$$

then from (4.13) we obtain that for any $K \geq 0$ (remembering that $\mathrm{E}\left[T_{\tau^0}\right] = \mathrm{E}\left[T\right]$)

$$\mathrm{E}\left[T_{\tau K+1}\right] = C_\tau + (1 - F(\tau))\mathrm{E}\left[T_{\tau K}\right], \tag{4.18}$$

and from (4.9) that for unbounded restarts:

$$\mathrm{E}\left[T_\tau\right] = \frac{C_\tau}{F(\tau)}. \tag{4.19}$$

Combining (4.18) and (4.19) it is easy to show that

$$\mathrm{E}\left[T_{\tau K+1}\right] < \mathrm{E}\left[T_{\tau K}\right] \iff \frac{C_\tau}{F(\tau)} < \mathrm{E}\left[T_{\tau K}\right] \iff \mathrm{E}\left[T_\tau\right] < \mathrm{E}\left[T_{\tau K}\right].$$

Since this holds for any $K \geq 0$, it follows that $\mathrm{E}\left[T_\tau\right] < \mathrm{E}\left[T\right]$, proving (4.16). To show that (4.17) holds, we derive:

$$\mathrm{E}\left[T - \tau \mid T > \tau\right] = \frac{\int_\tau^\infty (t - \tau)f(t)dt}{1 - F(\tau)}$$

$$= \frac{\int_\tau^\infty tf(t)dt - \tau(1 - F(\tau))}{1 - F(\tau)} = \frac{\mathrm{E}\left[T\right] - M_1(\tau)}{1 - F(\tau)} - \tau.$$

Then (4.17) follows using (4.9):

$$\mathrm{E}\left[T\right] + c < \mathrm{E}\left[T - \tau \mid T > \tau\right] \iff \mathrm{E}\left[T\right] + c < \frac{\mathrm{E}\left[T\right] - M_1(\tau)}{1 - F(\tau)} - \tau$$

$$\iff (1 - F(\tau))\mathrm{E}\left[T\right] < \mathrm{E}\left[T\right] - M_1(\tau) - (1 - F(\tau))(\tau + c)$$

$$\iff \mathrm{E}\left[T\right] > \frac{M_1(\tau)}{F(\tau)} + \frac{1 - F(\tau)}{F(\tau)}(\tau + c) \iff \mathrm{E}\left[T_\tau\right] < \mathrm{E}\left[T\right].$$

$\square$

Note that the above proof shows that the backward algorithm is a fixed-point iteration of the form

$$x_{K+1} = C_\tau + (1 - F(\tau))x_K,$$

(see (4.18)) with initial guess $x_0 = \mathrm{E}\left[T\right]$ and fixed-point solution $\mathrm{E}\left[T_\tau\right]$. The consequence of the first result of Theorem 4.6 is depicted in Fig. 4.7. The straight line is

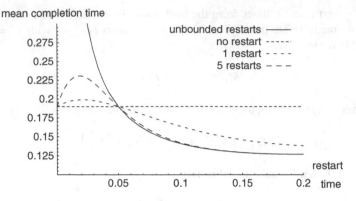

Fig. 4.7 Expected completion time for varying number of restarts

the expected completion time $E[T]$ without restarts, and the curve with the highest maxima and lowest minima is $E[T_\tau]$ for unbounded restarts. Because of Theorem 4.6, all curves improve the completion time over the same range of restart times, and the more restarts, the better. Similarly, when the completion time increases with restarts, fewer restarts are less detrimental for the completion time.

Another consequence of Theorem 4.6 is that if the completion time distribution is such that there exist restart times that improve expected completion time, as well as restart times that increase expected completion time, then there must also exist at least one point in which *all* curves cross, that is, it is immaterial if and how many restarts one executes. Figure 4.7 shows this, at $\tau \approx 0.05$.

We finally note that the results from Theorem 4.6 do not extend to higher moments. Figure 4.8 shows that there exist values for which one or two restarts improve the second moment, but unbounded restarts do not. There also is no point

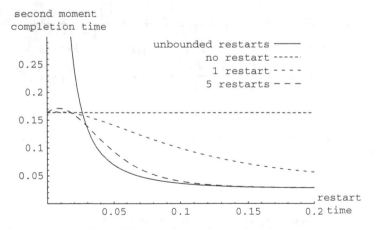

Fig. 4.8 Second moment of completion time for varying number of restarts

τ' in which any number of restarts provides the same completion time. The reason higher moments are more problematic is the same as why the backward algorithm does not work for higher moments: whether a restart time improves the completion time's higher moments is sensitive to the starting point of a restart interval. Repeated restarts may therefore not always keep improving the completion time's higher moments. Nevertheless, one can follow a similar conditional argument as in Sect. 3.2.1 to obtain the condition under which restart is beneficial for higher moments, namely $E\left[(T+c)^n\right] < E\left[(T-\tau)^n|T>\tau\right]$. However, from this we cannot conclude anything about the success for multiple or unbounded number of restarts.

4.3.2 Optimal Restart Times for Higher Moments

In Sect. 4.2 we formulated a mathematical description of models for moments of completion time under restart. We were able to derive a general description for all moments, the first moment as well as higher moments of completion time. However, optimising the moments with respect to the restart intervals can not be done for all moments similarly. Different algorithms are needed for higher moments than for the expected completion time. In the previous section we derived an algorithm for optimising the first moment of completion time under unbounded and finite many restarts. We provide in this section an efficient algorithm to determine the optimal time instants at which to initiate restarts, so that higher moments of completion time are minimised. As we showed in the previous section and in [163], the optimal restart strategy for the first moment can be determined in straightforward manner, both for finite and infinite number of allowed restarts. However, determining restart times that minimise higher moments of completion time is considerably more challenging, and requires an iterative approach to deal with the multiple dimensions of the optimisation problem. Our proposed algorithm leverages various expressions for the moments of completion time to simplify the minimisation problem. Because of this simplification, the algorithm outperforms more naive approaches by up to an order of magnitude (see Algorithm 4 in this section).

From our analysis it follows that it is typically not optimal to apply restarts at constant intervals when minimising higher moments, even if an infinite number of restarts is allowed. This is in contrast to the situation for the first moment, as we explained. We also provide insights into the characteristics of the optimal restart strategy through approximations under limiting conditions. It turns out that as long as enough restarts are allowed, one can use a first-moment approximation, with appropriate corrections for the first and last few restarts. Such approximations enable quick estimates for optimal restart times, and are therefore of practical importance.

When optimising the moments of completion time separately one finds that a restart time can be optimal for one moment but not for others, i.e. each moment has its respective optimal restart times. We will see that the cusp point (which minimises

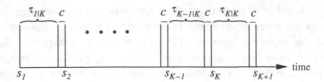

Fig. 4.9 Labelling restarts, total of K restarts

both 'reward' and 'risk') identified in [104] does not generally exist, for instance, not for the mixed hyper/hypo-exponential distribution which we use for illustration.

Optimising higher moments needs slightly heavier notation, since we have to always account for the total number K of restart intervals, which we therefore need to incorporate into the definition of moments of completion time. We use the expression for the density and distribution of the random variable for completion time under restart in slightly different form than (4.4) and (4.4) and therefore recall both of them here. Also, our numbering scheme must be augmented with the total number of restart intervals K, as shown in Fig. 4.9.

Remember the random variable T representing the completion time of a job without restarts, $f(t)$ its probability density function, and $F(t)$ its distribution. The total number of restarts is K, and the overhead associated with restarting is again c time units for each restart. The random variable T_K (with density $f_K(t)$ and distribution $F_K(t)$) represents the completion time with K restarts, where the restart intervals have length $\tau_{1|K}, \ldots, \tau_{K|K}$, (see Fig. 4.9). The k-th interval starts at time s_k, that is, $s_1 = 0$, $s_2 = \tau_{1|K} + c, \ldots, s_{K+1} = \sum_{k=1}^{K} \tau_{k|K} + Kc$.

Setting $\tau_{K+1|K} = \infty$ for notational purposes, the density function $f_K(t)$ and survival function $\bar{F}_K(t) = 1 - F_K(t)$ depend on which restart interval t falls in, as follows [163]:

$$
f_K(t) = \begin{cases} \prod\limits_{i=1}^{k-1} \bar{F}(\tau_{i|K}) f(t - s_k) & \text{if } s_k \leq t < s_k + \tau_{k|K}, \quad k = 1, \ldots, K+1 \\ 0 & \text{if } s_k + \tau_{k|K} \leq t < s_{k+1}, \quad k = 1, \ldots, K \end{cases}
\tag{4.20}
$$

$$
\bar{F}_K(t) = \begin{cases} \prod\limits_{i=1}^{k-1} \bar{F}(\tau_{i|K}) \bar{F}(t - s_k) & \text{if } s_k \leq t < s_k + \tau_{k|K}, \quad k = 1, \ldots, K+1 \\ \prod\limits_{i=1}^{k} \bar{F}(\tau_{i|K}) & \text{if } s_k + \tau_{k|K} \leq t < s_{k+1}, \quad k = 1, \ldots, K \end{cases}
$$

The N-th moment $\mathrm{E}\left[T_K^N\right]$, our metric of interest, is by definition:

$$
\mathrm{E}\left[T_K^N\right] = \int_0^\infty t^N f_K(t) dt = \sum_{k=1}^{K+1} \int_{s_k}^{s_k + \tau_{k|K}} t^N f_K(t) dt.
\tag{4.21}
$$

To find the restart times $\tau_{1|K}, \ldots, \tau_{K|K}$ that minimise $E[T_K^N]$, one could minimise (4.21) directly. It results in a K-dimensional minimisation problem that can be solved with off-the-shelf optimisation software. However, it is computationally expensive, since every new 'guess' for $\tau_{k|K}$ implies recomputing the integral term in (4.21) for all intervals $[s_l, s_l + \tau_{l|K})$ with $l \geq k$, to determine if the guess improves $E[T_K^N]$.

As an alternative, we can use an expression for $E[T_K^N]$ we derived in 4.2. We present this expression here in slightly different form, using the notation $\bar{k} = K - k + 1$. It is important to grasp the intuitive meaning of $\bar{k}$: where k is the number of restarts preceding and including the k-th, $\bar{k}$ is the number of restarts succeeding and including the k-th. We then recursively relate moments for $\bar{k}$ restarts with that for $\bar{k} - 1$ restarts by adding one restart before the existing $\bar{k} - 1$:

$$E\left[T_{\bar{k}}^N\right] = M_k[T^N] + \bar{F}(\tau_{k|K}) \sum_{n=0}^{N} \binom{N}{n} (\tau_{k|K} + c)^{N-n} E\left[T_{\bar{k}-1}^n\right], \quad (4.22)$$

where $M_k[T^N]$ denotes the 'partial moment,' defined for $\bar{k} = 1, \ldots, K$, as:

$$M_k[T^N] = \int_0^{\tau_{k|K}} t^N f(t) dt.$$

Instead of minimising (4.21) one can minimise (4.22). In this case, however, every new 'guess' for $\tau_{k|K}$ implies computing $E[T_K^N]$ 'all the way,' recursively calculating $E[T_l^N]$ for all values $l \geq k$, to determine if the guess improves $E[T_K^N]$. This also introduces much computational overhead. (In Fig. 4.14 we will see that minimising (4.22) is in fact slightly less expensive than minimising (4.21), at least for the discussed example.)

The main idea is not to minimise (4.21) or (4.22) directly, but instead extend to higher moments an idea that worked very well in [163] for the first moment and is carried out in Sect. 4.3.1.2. Utilising the recursion of (4.22), the backward optimisation algorithm sequentially determines the restart time $\tau_{k|K}$ that minimises $E[T_{\bar{k}}] = M_k[T] + \bar{F}(\tau_{k|K})(\tau_{k|K} + c + E[T_{\bar{k}-1}])$, for $\bar{k} = 1$ to K. Its correctness relies on the fact that the optimal restart time $\tau_{k|K}$ is independent of preceding restarts, as we discuss in Sect. 4.3.1.2 and in detail below. We named this algorithm the backward algorithm, since it determines the best restart times in reversed order, that is, first $\tau_{K|K}$, then $\tau_{K-1|K}$ until finally $\tau_{1|K}$. A single pass of K optimisations is guaranteed to provide the optimal restart times, which makes the backward algorithm computationally much more efficient than minimising either (4.21) or (4.22). (Figure 4.14 shows an improvement of approximately a factor 20.)

As we already pointed out earlier, the backward algorithm can not be applied to higher moments, because the optimal value of a restart interval depends on the restarts that precede it. To resolve this problem, we now prove that the optimal restart time at interval k depends on preceding restart times solely through the sum of these restart times, not the individual values. That is, the optimal $\tau_{k|K}$ depends on $\tau_{1|K}, \ldots, \tau_{k-1|K}$ only through the value of s_k. Based on this, we obtain a new expression (namely expression (4.23)), which we combine with (4.22) to simplify the optimisation task.

Theorem 4.7 *For any strictly positive* $k \leq K$, *let the first* $k - 1$ *restart times* $\tau_{1|K}, \ldots, \tau_{k-1|K}$ *be given. The last* $\bar{k}$ *restart times* $\tau_{k|K}, \ldots, \tau_{K|K}$ *minimise* $\mathrm{E}\left[T_K^N\right]$ *if and only if they minimise* $\mathrm{E}\left[(T_{\bar{k}} + s_k)^N\right]$ *(where we equate restart* $\tau_{k+i-1|K}$ *in* T_K *with* $\tau_{i|\bar{k}}$ *in* $T_{\bar{k}}$, *for* $i = 1, \ldots, \bar{k}$).

Proof First, by definition:

$$\mathrm{E}\left[T_K^N\right] = \int_0^{s_k} t^N f_K(t) dt + \int_{s_k}^{\infty} t^N f_K(t) dt.$$

Since the left most integral term does not depend on $\tau_{k|K}, \ldots, \tau_{K|K}$, the last $\bar{k}$ optimal restart times minimise $\mathrm{E}\left[T_K^N\right]$ if and only if they minimise $\int_{s_k}^{\infty} t^N f_K(t) dt$. If we equate $\tau_{k+i-1|K}$ with $\tau_{i|\bar{k}}$ for $i = 1, \ldots, \bar{k}$, we know from (4.20) that for any $t \geq 0$:

$$f_K(t + s_k) = \prod_{l=1}^{k-1} \bar{F}(\tau_{l|K}) f_{\bar{k}}(t).$$

This implies that:

$$\int_{s_k}^{\infty} t^N f_K(t) dt = \int_0^{\infty} (t + s_k)^N f_K(t + s_k) dt =$$

$$\prod_{l=1}^{k-1} \bar{F}(\tau_{l|K}) \int_0^{\infty} (t + s_k)^N f_{\bar{k}}(t) dt = \prod_{l=1}^{k-1} \bar{F}(\tau_{l|K}) \mathrm{E}\left[(T_{\bar{k}} + s_k)^N\right].$$

The product in this expression is independent of $\tau_{k|K}, \ldots, \tau_{K|K}$, and therefore minimising $\int_{s_k}^{\infty} t^N f_K(t) dt$ (and thus $\mathrm{E}\left[T_K^N\right]$) corresponds to minimising $\mathrm{E}\left[(T_{\bar{k}} + s_k)^N\right]$ (with $\tau_{k+i-1|K} = \tau_{i|\bar{k}}, i = 1, \ldots, \bar{k}$). □

Theorem 4.7 implies that for any k, $k = 1, \ldots, K$, determining the optimal restart time $\tau_{k|K}$ corresponds to minimising:

$$\mathrm{E}\left[(T_{\bar{k}} + s_k)^N\right] = \sum_{n=0}^{N} \binom{N}{n} s_k^{N-n} \mathrm{E}\left[T_{\bar{k}}^n\right], \tag{4.23}$$

where $\mathrm{E}\left[T_{\bar{k}}^n\right]$ obeys (4.22):

$$\mathrm{E}\left[T_{\bar{k}}^n\right] = M_k[T^n] + \bar{F}(\tau_{k|K}) \sum_{m=0}^{n} \binom{n}{m} (\tau_{k|K} + c)^{n-m} \mathrm{E}\left[T_{\bar{k}-1}^m\right]. \tag{4.24}$$

We are now in a position to show that for the first moment, the optimal value for $\tau_{k|K}$ does not depend on earlier restarts, not even through their sum s_k. For that reason, the backward algorithm works correctly for the first moment. We present this as a corollary of Theorem 4.7.

Corollary 4.8 *The restart times $\tau_{k|K}, \ldots, \tau_{K|K}$ minimise $E[T_K]$ if and only if they minimise $E\left[T_{\bar{k}}\right]$ (with $\tau_{k+i-1|K} = \tau_{i|\bar{k}}$, $i = 1, \ldots, \bar{k}$).*

Proof This corollary follows from the fact that Theorem 4.7 states for N=1 that minimising $E[T_K]$ corresponds to minimising $E\left[T_{\bar{k}} + s_k\right]$. Obviously, $E\left[T_{\bar{k}} + s_k\right] = E\left[T_{\bar{k}}\right] + s_k$, and since s_k is a constant, it does not influence the optimisation solution. Therefore minimising $E[T_K]$ corresponds to minimising $E\left[T_{\bar{k}}\right]$. □

Minimising the second or third moment of completion time using (4.23) and (4.24) saves about 70% computation time for our example, as can be seen in Fig. 4.14 in Sect. 4.3.2.2. The reason for this speed-up is that with every 'guess' of $\tau_{k|K}$ when minimising (4.23) using (4.24), we only recompute $E\left[T_{\bar{k}}^n\right]$. That is, the algorithm neither requires to recompute integral terms for all values $l \geq k$ (as in (4.21)), nor terms $E\left[T_l^N\right]$ for all values $l \geq k$ (as in (4.22)). Effectively, we have isolated the optimisation of the k-th restart time from interference with the other restart intervals.

The resulting optimisation algorithm is given as Algorithm 4. Contrary to the backward optimisation algorithm for the first moment it does not terminate in K steps, but requires to iterate until convergence (of either $E\left[T_K^N\right]$ or the restart times). One can apply generic approaches to decide which restart time τ_k to optimise at each iteration (such as the method of steepest descent). However, we propose three particular ways, which try to leverage the structure of the problem: backward, forward and alternating.

Algorithm 4 (Backward, Forward and Alternating Optimisation)

```
Input constants N and K;
Input boolean alternating;
Set either boolean forward or backward to TRUE;
Determine τ∞ that minimises E[T∞];
For n = 1 to N {
        Compute and Set E[T₀ⁿ] (moments without restarts);
        For k = K to 1
                Initialise E[T_k̄ⁿ] using (4.24) with τk|K = τ∞;
}
While( not converged ) Do {
        If( backward ) then {
                For k = K to 1 {
                        Find τk|K that minimises (4.23), using (4.24) for E[T_k̄ⁿ];
                        For n = 1 to N
                                Update E[T_k̄ⁿ] using (4.24) with new value of τk|K;
                }
        If( forward ) then {
                For k = 1 to K
```

```
                Find τ_{k|K} that minimises (4.23), using (4.24) for E[T_k^n];
                Update s_{k+1} with new value of τ_{k|K};
            }
        For k = K to 1 {
            For n = 1 to N
                Update E[T_k^n] using (4.24) with new value of τ_{k|K};
            }
        }
        If( alternating ) then swap backward and forward
    }
    Return τ_{1|K},...,τ_{K|K};
```

Each minimisation step in the algorithm can be carried out with any desired general-purpose optimisation routine. Also, note that at initialisation, $E[T_\infty]$ can be minimised using the expression $E[T_\infty] = (M_1[T] + \bar{F}(\tau_\infty)(\tau_\infty + c))/F(\tau_\infty)$ derived in Sect. 4.2.1. The reason to initialise the algorithm with τ_∞ will become apparent when discussing the bulk approximation in the next section.

4.3.2.1 Characteristics of Optimal Restart Times

We applied our algorithm to the case that the completion time T has a lognormal distribution, with parameters $\mu = -2.31$ and $\sigma = 0.97$.[3] We determine $K = 15$ restart times that minimise the first, second and third moment of the completion time. These restart times (with an interpolating curve) are shown in Fig. 4.10. The figure also shows τ_∞, which is the starting solution set at the initialisation step in Algorithm 4.

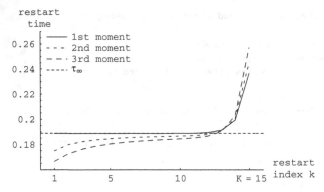

Fig. 4.10 Optimal restart times, with respect to the moments $E[T_{15}]$, $E[T_{15}^2]$ and $E[T_{15}^3]$, respectively

---

[3] There is no particular significance to the chosen parameter values. They happen to be the parameters of a lognormal fit for experimental data of HTTP GET completion times [128].

Figure 4.10 indicates that when minimising the first moment, the optimal restart time $\tau_{k|K}$ monotonically converges when k gets smaller, to a single optimum τ_∞, provided K is large enough. In fact, as we observed in Sect. 4.3.1.3, the backward algorithm is a fixed-point algorithm, with associated convergence properties. This also implies that if an infinite number of restarts is allowed, a constant restart time is optimal, as has been observed in [5, 99].

The convergence behaviour for higher moments is not as straightforward, as seen in Fig. 4.10. Nevertheless, there are some interesting insights to be gained from explaining the more intricate convergence patterns.

The key observation is that if the number of restarts increases, the dominant term when minimising (4.23) involves only the first moment. Therefore the restart times that minimise the first moment are a good strategy for higher moments as well (provided K is large, and not considering the first and last few restart times). To make this more precise, assume that $k \to \infty, k \leq K$, in which case, apart from pathological cases, it must be that $s_k \to \infty$. This allows us to approximate expression (4.23) by the first two terms of its sum:

$$\lim_{k \to \infty} \mathrm{E}\left[(T_{\bar{k}} + s_k)^N\right] \approx s_k^N + N s_k^{N-1} \mathrm{E}\left[T_{\bar{k}}\right]. \tag{4.25}$$

Since s_k^N and $N s_K^{N-1}$ are constants, finding the restart times that minimise (4.23) is approximately equal to finding the restart times that minimise the first moment $\mathrm{E}\left[T_{\bar{k}}\right]$. Based on this, we introduce three limiting cases, namely at the right boundary ($\tau_{k|K}$ for $k \to \infty$, and $\bar{k} \downarrow 1$), middle or 'bulk' ($\tau_{k|K}$ for $k \to \infty$ and $\bar{k} \to \infty$), and left boundary ($\tau_{k|K}$ for $k \downarrow 1$ and $\bar{k} \to \infty$). Figure 4.11 illustrates the main results.

Right boundary approximation. For $k \to \infty$ and $\bar{k} = 1$ the first-moment approximation of $\tau_{k|K}$ corresponds to finding the restart time that minimises $\mathrm{E}[T_1]$ (only one restart allowed). Figure 4.11 shows the right approximation and we see that for $K = 30$ it is reasonable but not exceptionally close to the actual optimal restart

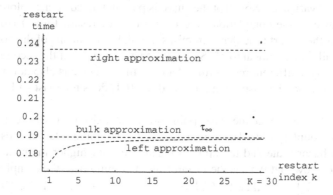

Fig. 4.11 The dots give restart times that minimise the second moment $\mathrm{E}\left[T_{30}^2\right]$, the dashed lines are the approximations, as labelled

(given by the dot). We can extend the approximations to value $\bar{k} = 2, 3, \ldots$, which results in restart times identical to those shown in Fig. 4.10 (the curve for the first moment).

Bulk approximation τ_∞. At the 'bulk,' or the middle of the pack, we get a limiting result if both k and $\bar{k}$ go to infinity. The approximation using (4.25) results in optimising $\mathrm{E}\,[T_\infty]$, i.e., in restart times equal to τ_∞. In Fig. 4.11, the bulk approximation τ_∞ is indeed close to the optimal restart times. This also explains why we chose τ_∞ during initialisation in Algorithm 4: it is close to optimal for the bulk of restarts.

Left boundary approximation. At the left boundary, we can not simply apply the approximation suggested by equation (4.25), because k does not tend to infinity. However, we can obtain an approximation for $\tau_{k|K}$ with $k \downarrow 1$ and $\bar{k} \to \infty$, by assuming that the completion time T is distributed as T_∞ with restart interval τ_∞. This approximation is remarkably close, as seen from Fig. 4.11. In fact, other experiments indicate that the left boundary approximation is very close irrespective of the value of K. This implies that if we allow an infinite number of restarts, we can use the left boundary approximation to determine early restarts, until it is close enough to the bulk approximation (which we would use from then on).

In conclusion, we find for the first moment of completion time that the optimal restart strategy is a constant restart time for all restarts, provided we allow an infinite number of restarts. If only a finite number of K restarts is allowed, we can optimise these using the backward algorithm, which terminates in K steps. When we consider higher moments, a constant restart time is typically not optimal, not even if we allow infinitely many restarts. Instead, we need the backward/forward iterative algorithm to compute optimal restart times for the finite case, and use the bulk and left boundary approximation for the case with infinitely many restarts.

Cusp Point. In [104] the authors point to the existence of a 'cusp point', in which both the expected completion time and its variance are minimised. In terms of [104] 'reward' as well as 'risk' are then optimised jointly. Figure 2 in [104] suggests that such a cusp point exists; we have redone this in Fig. 4.12 for our example, and indeed two curves seem to come together at a cusp point where it reaches the minimum for both mean and variance. (Note that the curve is parameterised over τ, plotting mean versus variance of the completion time for a range of restart times.) However, it turns out that the restart time that minimises the higher moments of the completion time is typically not identical to τ^*. Since it is easy to see that if the second moment is minimised by a different restart time than τ^*, the variance is also not minimised in τ^*, it follows that the cusp point identified in [104] does not exist–at least, not in general.

One way to show that the cusp point does not exist is to derive for higher moments the counterpart to Theorem 4.4 on p. 63, so that a relation is established between the hazard rate and the optimal restart time for higher moments. Then it is possible to show that if τ^* is the restart time that minimises the completion time for $\mathrm{E}\,[T_\tau], \ldots, \mathrm{E}\,[T_\tau^N]$, then $\mathrm{E}\,[T_{\tau^*}^n] = n!(\frac{1-F(\tau^*)}{f(\tau^*)})^n$, for $n = 1, \ldots, N$. We can certainly construct probability distributions with partial moments $M_n(\tau)$ such that

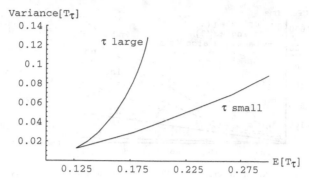

Fig. 4.12 Mean and variance of the completion time, parameterised by restart time τ, as Fig. 2 of [104]

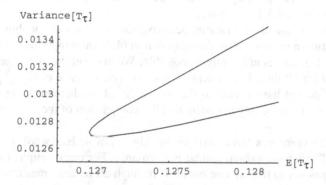

Fig. 4.13 As Fig. 4.12, zoomed in at 'cusp point'

this relation holds when filling in (4.6), but in general the relation will not hold true. Instead of providing the proof for this negative result, we demonstrate numerically that the cusp point does not exist for our running example. Figure 4.13 zooms in at the 'cusp point' of Fig. 4.12 and demonstrates that there is no point that minimises the curve with respect to both the x and y-axis (mean and variance of completion time). In particular, for our example, the minimum expected completion time is for restart time $\tau^* = 0.198$, for the second moment the minimum is for $\tau = 0.192$, and for the variance the minimum is reached at $\tau = 0.188$.

4.3.2.2 Computational Effort

In Fig. 4.14 we plot the time used for three different methods: Algorithm 4 (backward), minimising expression (4.21), and minimising expression (4.22). In all cases we applied default minimisation algorithms in Mathematica to carry out the respective optimisation steps. Algorithm 4 outperforms the other approaches. For the first moment, the speed up is an order of magnitude (about a factor 20), which finds its

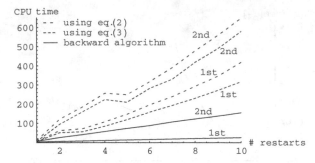

Fig. 4.14 CPU time used for different algorithms, applied to minimise first as well as second moment

explanation in assured convergence in K steps of the backward algorithm. For the higher moments, the speed up is about a factor 3 or 4. Apparently, the arguments put forward in Sect. 4.3.2 hold correct.

We note that we tuned our Mathematica program to the best of our abilities, memorising in-between results so that the recursion in (4.22) and repetitive computation in (4.21) are handled as efficiently as possible. We also set the convergence criterion identical for all three experiments, (based on convergence of $\mathrm{E}\left[T_K^N\right]$). Hence, although we do not have access to the 'internals' of Mathematica's optimisation algorithm, we are reasonably confident that the comparison of the three approaches is fair.

Figure 4.15 compares three versions of Algorithm 4: backward, forward and alternating. These three exhibit similar performance. For our example, the forward algorithm turns out to require one pass less through all restart times than the other two algorithms, and hence it takes less CPU time.[4]

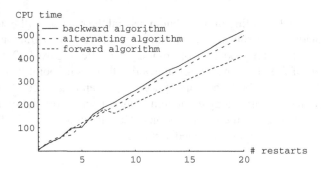

Fig. 4.15 CPU time used by three versions of algorithm

---

[4] Note that because of the workings of the Mathematica optimisation algorithm, the comparison in Fig. 4.14 had to be based on convergence of $\mathrm{E}\left[T_K^N\right]$ as stopping criterion, while we were able to base the results for the backward/forward algorithm in Fig. 4.15 on convergence of restart times,

Typically, we require not more than five passes through the K restart times, irrespective of the value of K. Studying the complexity of our Mathematica implementation, it turns out that running the optimisation routine is the computationally most expensive part: at step k, optimisation of $\tau_{k|K}$ takes an order of magnitude more time than the computation that updates $\mathrm{E}\left[T_{\tilde{k}}^{N}\right]$. Algorithm 4 uses backward and/or forward traversal through the K restart times to compute $\mathrm{E}\left[T_{\tilde{k}}^{N}\right]$ efficiently, but following the above reasoning, it may be more important to decrease the number of calls to the optimisation routine. Algorithm 4 may therefore be further improved by choosing the order in which to optimise restart times based on criteria such as steepest descent. This requires more experimentation.

4.4 Case Study: Optimising Expected Completion Time in Web Services Reliable Messaging

In this section we will study the performance and potential of our Algorithm 1 on p. 58 for the computation of the optimal restart time to optimise the first moment of completion time if there are infinitely many restarts allowed. The results of this case-study have been published in [129].

We apply the algorithm in the setting of Web Services Reliable Messaging (WSRM) where a retransmission mechanism is needed to guarantee reliable data transmission. Several attempts of defining an appropriate reliability standard have converged in WSRM, which provides a framework to deliver messages 'reliably between distributed applications in the presence of software component, system, or network failures' [10]. Of the four delivery assurances specified in [10], both 'at least once' and 'exactly once' necessitate the retransmission of lost messages. While the standard describes positive and negative acknowledgements to determine the message transmission status, it does not provide details on the preferred waiting time (for a positive acknowledgement) until resending a message. Although exponential backoff is mentioned as one way to adjust the retransmission interval, the issue is effectively left open.

In this section we illustrate how the restart model may be a suitable candidate to compute the waiting time until resending a message in WSRM. We compare the timeout computation obtained from the restart model with three other options: exponential backoff, constant intervals and the Jacobson-Karn algorithm that is commonly used in TCP implementations. The Jacobson-Karn algorithm has been introduced in Sect. 3.1.4 on p. 40.

We experimentally investigate the influence of timeout strategies on the performance of, and overhead introduced by, WSRM. In particular, we analyse representative algorithms for four classes of restart oracles (as explained in Sect. 4.4.2). We

---

a stricter criterion. This explains the higher CPU usage for the backward algorithm in Fig. 4.15 compared to Fig. 4.14.

see that the more complex the behaviour of the underlying network and system, the more it pays off to utilise strategies that adapt the time-out value based on observed data.

There are two important aspects to the evaluation of WSRM timeout strategies: the optimal strategy as decidable at the level of WSRM and the interaction with reliability mechanisms at lower layers, in particular TCP. Especially the latter is extremely difficult to track. Therefore, we have created a fault injection environment for the analysis of WSRM, in which the transmission of IP packets can be interrupted. This allows us to mimic scenarios that from the perspective of the WSRM layer are representative for a variety of network, protocol and system failure behaviours and represent situations where retransmission is necessary for reliable data transfer. Since we inject faults at the IP level, we can experimentally evaluate the complex interaction between TCP and WSRM reliability mechanisms.

With restart, there always exists a tradeoff between fairness and timeliness. Whereas shorter intervals help to overcome faults faster, the resulting more frequent restarts consume more resources. Longer timeouts are more conservative, but also slow down recovery and thus message end-to-end delay. For all four timeout oracles we evaluate their application with respect to the tradeoff between fairness and timeliness.

There exists some work related to our analysis of WSRM timeout mechanisms using web service fault injection for WSRM performance. Looker and Xiu [97] present a framework to inject faults using a modified SOAP layer. This allows one to explore the impact of specific web service failures, but this approach does not yield an understanding of the effects fault handling in the lower layers has on higher layers. At lower levels of the stack, dependability has been studied more extensively. For instance, Allman and Paxson [4] explore how different parameters of Retransmission Timeout (RTO) algorithms affect fault recovery in TCP. We study several oracles in regard to the general tradeoff between fairness and timeliness on a higher level. We inject faults in the IP layer and thus elicit fault-handling in all layers beneath WSRM.

Figure 4.16 depicts a basic WSRM deployment. Resting beneath the application and atop a stack of further layers,[5] the WSRM component provides reliable message transfer between the application's endpoints. The direction of message flows is reflected in the distinction between a 'WSRM Source' and a 'WSRM Destination': the sending party is the source, while the recipient is the destination. To ensure transmission, the source keeps resending each message until it receives an acknowledgement from the destination.

There are several complex layers beneath WSRM. The SOAP layer offers an abstract medium. SOAP implementations (e.g., Axis [149]) then use lower-level protocols as SOAP transports to send and receive messages. As these transports

---

[5] Adhering to SOAP terminology, we do not call HTTP an application, but rather a transport for the layers above it.

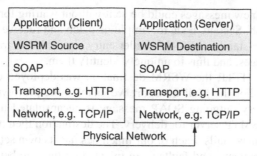

Fig. 4.16 A sample WSRM setting

are often application level protocols themselves (e.g., HTTP is an application level protocol in the TCP/IP stack), they in turn utilise lower network protocol layers.

Each higher layer's reliability is influenced by faults in the layers beneath it. These faults may eventually lead to failing message transfers. On the other hand, some lower layers offer reliability as well. TCP, for instance, resends unacknowledged IP packets to provide reliable connections over unreliable IP networks. However, fault handling at any point in the stack does not necessarily guarantee reliability at the uppermost level. Furthermore, fault handling mechanisms may lead to undesirable timing behaviour in upper layers.

To elicit fault handling in the stack (including WSRM), we inject faults into the IP layer. By injecting faults at this low level we are able to investigate the interaction of TCP and WSRM reliability mechanisms and hence study the effects of different timeout algorithms in practice. The injected faults correspond to IP packet loss, and may lead to delays and losses in message transfer. From the perspective of the WSRM layer, these effects may be representative for a wide variety of failure scenarios, including server hardware and software as well as network failures.

Figure 4.17 shows the basic setting for the experiments.

Transmission dependability was provided by the WSRM implementation Sandesha 1.0 RC1 [150] with in-memory storage of messages and some modifications, as follows. First, Sandesha and its SOAP engine Axis 1.2RC2 [149] were

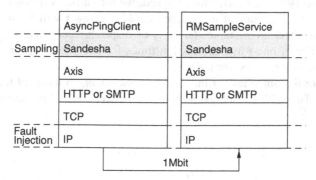

Fig. 4.17 Basic experiment setup

de-coupled so that new messages could be sent without waiting for previous SOAP invocations to finish. Second, code to sample dispatch and receive timestamps was added. Third, a non-standard WSRM header entry was added to track transmission numbers for messages, and thus to uniquely identify them.

For SOAP over HTTP, the WSRM destination was deployed on a Tomcat 4.1 application server. For the Mail transport we enabled the prototypical implementation in Axis. Every outgoing SOAP message is encapsulated in an E-Mail and handed over to an SMTP server for delivery. The destination then periodically polls a POP3 server for new mails. Each of our machines has its own set of SMTP/POP3 servers (Postfix/Qpopper), and faults were injected on the link between the SMTP servers. Mail servers retry delivery after non-fatal faults. Deferred mails are stored in a queue and assigned a timeout for retransmission. If the fault was caused by the destination (e.g., a failed connection), no delivery to that host will be attempted for some time. The queue is periodically checked for mails to be retransmitted. In order to reduce load, and since E-Mail is not considered time-critical, all of these intervals are usually large. When delivery time becomes more important, lower timeouts seem more appropriate. We set the queue check interval and the minimum and maximum waiting times to 60, 60 and 120 s, respectively, and polled the POP3 server with an interval of 6 s.

Two Linux machines, connected by a 100Mbit network and kept synchronised through NTP provided the physical environment. Paxson [119] notes that NTP is designed for long-term synchronisation and thus may not be sufficiently accurate in small timescales. We tried to minimise these issues by explicitly synchronising one machine to the other before each experiment run and disabling synchronisation during the runs. On this basis we emulated a 1Mbit/1Mbit up/downstream network, which seems to us a minimum requirement for most typical applications using web services. Faults typical of poor network conditions were injected into the IP layer using a combination of Linux's NetEm queueing discipline and IP firewalling facilities.

4.4.1 Metrics for the Fairness-Timeliness tradeoff

To describe precisely the metrics we analysed, we formalise as follows. Each message $m_i, i = 1, \ldots M$ out of a sequence of M messages will be sent $n_i \geq 1$ times, and thus there are n_i tuples (s_{ij}, r_{ij}) of dispatch/receipt times $s_{ij}, r_{ij} \in \mathbb{R}$ for each m_i. The n_i one-way transmission times for every message are defined as $t_{ij} = r_{ij} - s_{ij}$, with $t_{ij} = \infty$ for failures.

We measure the timeliness of individual messages m_i in terms of the Effective Transmission Time (ETT), defined as the time between m_i's first dispatch and its first arrival:

$$r_i^* := \min\{r_{ij}\}$$
$$ETT_i := r_i^* - s_{i1}.$$

The following line of thought yields a generic measure of fairness. m_i is sent $n_i \geq 1$ times. A number of these transmissions may fail, but at time $r_i^* < \infty$ the first one (with index k_i^*) is successful. Now while other transmissions initiated before r_i^*, but completed later might have taken longer than k_i^*, they were not strictly necessary. Additionally, every retransmission after r_i^* is clearly obsolete. Thus, the only unavoidable retransmissions are those that started before r_i^*, but failed, and k_i^*. With every other transmission the sender consumed resources it did not really need. We term this behaviour Unnecessary Resource Consumption (URC)[6] The computation of URC resulting from this definition is slightly pessimistic, as it may deem some lost transmissions unnecessarily sent if there were several 'first' successful transmissions (i.e. if $|MIN| > 1$). However, this case is very unlikely. Also, due to the costs associated with URC, a pessimistic metric seems appropriate.

$$k_i^* := \operatorname*{argmin}_j \{r_{ij}\}$$

$$URC_i := n_i - |\{r_{ij} = \infty | j < k_i^*\}| - 1$$

Consuming resources unnecessarily is unfair towards other users of the same resources. High URC thus indicates unfair behaviour, and low URC indicates more fair behaviour. $URC_i = 0$ corresponds to the optimal case. Since URC is independent of the nature of the resource in question (e.g., network bandwidth, processing power) and of assumptions regarding other parties accessing it (i.e., those to be fair towards), it allows us to compare fairness across many applications. It is limited in that it centers on individual messages and does not show aggregate load. The latter, however, is usually determined by the application's sending rate, and therefore out of scope for restart oracles.

4.4.2 Oracles for Restart

In our experiments, we compared time-out oracles based on different algorithms: fixed, fixed with back-offs, dynamic intervals using Jacobson/Karn and dynamic intervals using Algorithm 1 on p. 58. These respective oracles are representatives of different classes of timeout oracles. We explain the classification and provide the details of the oracles in this subsection.

In short, oracles compute timeouts for messages using an algorithm. An oracle may potentially combine several timeout algorithms. For example, an oracle might choose between a fair and a fast algorithm based on additional system state data.

---

[6] Where $\operatorname*{argmin}_i \{x_{i\in \mathbb{I}}\} := \min \mathbb{I}_{\mathrm{Min}}$ with

$$\mathbb{I}_{\mathrm{Min}} := \{i \in \mathbb{I} : x_i \in Min\}$$
$$Min := \min \{x_i : i \in \mathbb{I}\} .$$

The oracle mechanism is an almost non-intrusive way to improve performance. It only changes retransmission intervals in the sender, hence it enhances a reliability mechanism already present in virtually any communication system.

Upon receipt of a positive message acknowledgement at time t the sender can determine the *transmission status* and estimate the *transmission time*. (Whereas from a negative acknowledgement it can only infer the status.)

Two aspects about estimating transmission time in this way are noteworthy. First, this notion of transmission time corresponds to the complete Round-Trip Time (RTT), rather than the one-way time from sender to receiver, and secondly, if acknowledgements do not carry additional information to assign them to a particular transmission, a 'retransmission ambiguity' [79] may reduce accuracy of these samples.

Recall that each message $m_i, i = 1, \ldots M$ out of a sequence of M messages will be sent $n_i \geq 1$ times, and thus there are n_i tuples (s_{ij}, r_{ij}) of dispatch/receipt times for each message. The n_i transmission times for every message then are $t_{ij} = r_{ij} - s_{ij}$, with $t_{ij} = \infty$ for lost messages. Without additional means, the sender cannot measure r_{ij}. Instead, it guesses the transmission time from positive acknowledgements. The sender observes

$$a_{ij} := \begin{cases} z & \text{an acknowledgement for } m_i \text{ arrived at time } z \\ \infty & \text{else,} \end{cases}$$

the arrival times of acknowledgements, from which it infers that the message has reached its destination some time before a_{ij}. It thus estimates t_{ij} by $\hat{t}_{ij} < a_{ij}$, e.g., by the RTT $\hat{t}_{ij}^{RTT} := a_{ij} - s_{ij}$. We distinguish the four types of oracles by how they utilise this data.

4.4.2.1 Oracles using Fixed Intervals

A timeout τ is given at compilation time or startup, and every τ time units the sender checks whether an acknowledgement for m_i has arrived, and resends m_i otherwise. I.e., if m_i has been sent at time T_i, the sender will look for an acknowledgement at $T_i + \tau, T_i + 2\tau, T_i + 3\tau, \ldots$.

Performance of a fixed intervals oracle hinges on whether τ is consistent with actual medium characteristics. In extremis, a timeout lower than the RTT leads to retransmissions of every message, and therefore a fixed interval will rarely be used in practical systems.

4.4.2.2 Oracles using Growing Intervals

These augment Fixed Intervals by additionally tracking oracle decisions. That is, if a timeout τ_{ij} for m_i expires, they assume τ_{ij} to have been too short and increase it by a substantial amount, e.g., exponentially:

$$\tau_{i1} = \tau$$
$$\tau_{i\,(j+1)} = 2 \cdot \tau_{ij}.$$

While this reduces unnecessary retransmissions, fairness does still depend on the initial timeout. Furthermore, any such approach hurts timeliness, which is most pronounced in scenarios where messages are sent or consumed synchronously.

4.4.2.3 Oracles using Basic Adaptive Intervals

With these oracles, the timeout is allowed to shrink as well as to grow. In essence, they parametrise an assumed probability distribution F_c for RTTs based on estimates of the transmission time, and base τ for all m_i on F_c. Because this global τ is usually closer to actual round trip times than one set beforehand, adaptive oracles can avoid unnecessary retransmissions and are able to detect loss in a timely manner. Their timeliness may be further improved by their potential to also detect (and restart) exceptionally slow invocations. However, their performance ultimately depends on (i) the similarity between the assumed and the actual distribution, (ii) the accuracy of RTT measurements, and notably (iii) the impact of the retransmission ambiguity as a factor in the accuracy of RTT measurements. Perhaps the best known example is the TCP RTO computation which keeps track of the mean and variance of RTTs using successive samples $\hat{t}_{ij}^l$, $l = 1, 2, \ldots$. After the smoothed round-trip time (SRTT) was initialised with the first observed RTT, subsequent $SRTT_{l+1}$ are computed using

$$SRTT_{l+1} := (1 - \alpha_1)SRTT_l + \alpha_1 \cdot \hat{t}_{ij}^{l+1}.$$

The variance measure

$$RTTVAR_1 = \frac{1}{2}SRTT_1$$
$$RTTVAR_{l+1} := (1 - \alpha_2) \cdot RTTVAR_l +$$
$$+\alpha_2 \cdot |SRTT_{l+1} - \hat{t}_{ij}^{l+1}|,$$

was introduced by Jacobson to improve similarity (i). In regard to the retransmission ambiguity, Karn and Partridge proposed to only use fresh samples $\hat{t}_{i1}$ from acknowledgements for which $n_i = 1$, and to back off exponentially upon encountering a timeout [79]. The retransmission timeout (RTO) is then computed as

$$\tau := RTO_{l+1} := \begin{cases} 2 \cdot RTO_l & \text{if } RTO_l \text{ expired} \\ SRTT_l + k \cdot RTTVAR_l & \\ \quad \text{if a new } \hat{t}_{i1}^{l+1} \text{ is available,} \end{cases} \tag{4.26}$$

where k is a flexible parameter, typically set to 4 [4], and starts with $RTO_0 = 3\,\text{s}$. Our basic adaptive intervals oracle (Jacobson/Karn) implements (4.26), and thus follows the notion of combining dependability and fast fault-recovery with conservative medium usage.

4.4.2.4 Oracles using Advanced Adaptive Intervals

The restart model discussed in this chapter starts from the premise that the completion times T of a task (e.g., the RTT for message transmission) are drawn independently from the same known probability density function $f(\hat{t}_{ij})$. Provided the distribution function $F(\hat{t}_{ij})$, or an estimate $\hat{F}(\hat{t}_{ij})$ thereof is known Algorithm 1 on p. 58 can be used to compute the optimal restart time. Since timeout selection is based on a comparison of anticipated transmission times, and because $\hat{F}(\hat{t}_{ij})$ can be learned from observations, oracles employing this approach potentially improve transmission times in many scenarios, even when the distribution for T is not known beforehand. Shortcomings lie in an inherent assumption that restart does preempt previous attempts, and, as with the Basic Adaptive Intervals, in the sensitivity to inaccuracies in measurements or estimates of the transmission time of message m_i, $\hat{t}_{ij}$. If the former is not true, restart may lead to overload. We therefore applied Karn's and Karn/Partridge's RTO algorithm modifications (viz. exponential backoff and ignorance towards samples from retransmissions) to the basic oracle to enhance its fairness.

The online algorithm in Appendix A to [128] learns $\hat{F}(\hat{t}_{ij})$ by building a histogram. It divides the range of observed transmission times $\hat{t}_{ij} < \tau_{max}$ into H buckets $k = 1, \ldots, H$ of size $h = \tau_{max}/H$. To find the optimal timeout, it computes estimates $\hat{E}_{\tau_k}$ of the expected transmission time for restart after each of the intervals and then selects the interval with the lowest $\hat{E}_{\tau_k}$. Hence, all intervals $\tau_k = k \cdot h$ are candidate retry times. To account for known constant delays directly associated with restart, a cost value c can be set. The estimate

$$\hat{E}_{\tau_k} = \frac{\hat{M}(\tau_k)}{\hat{F}(\tau_k)} + \frac{1 - \hat{F}(\tau_k)}{\hat{F}(\tau_k)} \cdot (\tau_k + c) \tag{4.27}$$

is obtained using the average return times M_k and number of samples N_k within the intervals $[(k-1) \cdot h, k \cdot h)$. If we label all known observations $\hat{t}_{ij}$ in bucket k $t_1^k, \ldots t_{N_k}^k$, M_k is estimated by

$$\hat{M}_k = \frac{1}{N_k} \sum_{l=1}^{N_k} t_l^k. \tag{4.28}$$

Using the number N_k of observations that take at least τ_k time units, the estimators of the distribution function of the job completion time and of its first partial moment evaluate to:

$$\hat{F}(\tau_k) = \frac{\sum_{l=1}^k N_l}{\sum_{l=1}^H N_l + N_{\tau_{max}}}$$

and

$$\hat{M}(\tau_k) = \frac{\sum_{l=1}^k N_l \cdot \hat{M}_l}{\sum_{l=1}^k N_l}.$$

The global timeout τ is chosen to be the optimal τ_k, i.e., the one that minimises the expected RTT in (4.27):

$$\tau := \underset{\tau_k}{\operatorname{argmin}} \, \hat{E}_{\tau_k}.$$

We label this oracle QEST, because of the venue of the original publication of the algorithm.

4.4.3 Results

We study two sets of scenarios. In the first one, an otherwise perfect network connection temporarily exhibits 100% packet loss of variable duration (10 and 60 s) after an initial 60 s warm-up period. In the second, network conditions remain stable throughout the experiment, with several levels of random packet loss enforced on outgoing packets on both sides. Based on Paxson's observation that average packet loss on the Internet may reach up to 5.2% [118], we inject loss at 0, 1 and 5% in each direction (resulting in an effective loss of 0, 2 and 10%, respectively, for bidirectional communication). Table 4.2 shows the parameter values for all oracles in our experiments. In addition to these oracles, we also conduct a series of experiments without restart.

Table 4.2 Oracle parameters. We chose longer initial timeouts for the Mail transport to account for the larger ETT observed without restart. Since static oracles cannot adapt, we added an extra safety margin to their timeouts. Parameters for the Jacobson/Karn Oracle were set based on [4]

	Fixed Intervals	Exp. Backoff	Jacobson/Karn	QEST Oracle
HTTP Transport				
	$\tau_{ij} := 4\,\mathrm{s}$	$\tau_i = 4\,\mathrm{s}, 8\,\mathrm{s},$ $16\,\mathrm{s}\ldots$	$\tau := 4\,\mathrm{s}$ $\alpha_1 = 1/8$ $\alpha_2 = 1/4$ $k = 4$	$\tau := 4\,\mathrm{s}$ $\tau_{max} := 60\,\mathrm{s}$ $H = 1000$ $c = 0\,\mathrm{s}$
Mail Transport				
	$\tau_{ij} := 14\,\mathrm{s}$	$\tau_i = 14\,\mathrm{s}, 28\,\mathrm{s},$ $56\,\mathrm{s}, \ldots$	$\tau := 10\,\mathrm{s}$ $\alpha_1 = 1/8$ $\alpha_2 = 1/4$ $k = 4$	$\tau := 10\,\mathrm{s}$ $\tau_{max} := 60\,\mathrm{s}$ $H = 1000$ $c = 0\,\mathrm{s}$

Table 4.3 Result summary (HTTP Transport). ETTs given in seconds, with 95% confidence intervals. ETT and URC for 'No Restart' in the last two scenarios are omitted because of message loss

	No Restart	Fixed Int.	Exp. Backoff	Jac./Karn	QEST
No Faults					
avg. ETT	0.17 ± 0.01	0.16 ± 0.00	0.18 ± 0.00	0.18 ± 0.00	0.16 ± 0.00
avg. URC	0.00 ± 0.00	0.00 ± 0.00	0.00 ± 0.00	0.03 ± 0.00	0.03 ± 0.00
10 s Disruption					
avg. ETT	0.27 ± 0.01	0.22 ± 0.01	0.23 ± 0.01	0.23 ± 0.01	0.27 ± 0.01
avg. URC	0.00 ± 0.00	0.01 ± 0.00	0.01 ± 0.00	0.04 ± 0.00	0.04 ± 0.00
60 s Disruption					
avg. ETT	3.03 ± 0.18	2.20 ± 0.12	2.12 ± 0.12	2.34 ± 0.13	2.38 ± 0.13
avg. URC	0.00 ± 0.00	0.42 ± 0.03	0.14 ± 0.01	0.06 ± 0.00	0.07 ± 0.00
2% Packet Loss					
avg. ETT	–	0.76 ± 0.02	0.77 ± 0.02	0.79 ± 0.02	0.73 ± 0.01
avg. URC	–	0.07 ± 0.00	0.08 ± 0.00	0.18 ± 0.01	0.34 ± 0.01
10% Packet Loss					
avg. ETT	–	3.83 ± 0.06	4.57 ± 0.10	4.78 ± 0.08	2.87 ± 0.04
avg. URC	–	0.25 ± 0.01	0.22 ± 0.01	0.25 ± 0.01	0.33 ± 0.01

Care has been taken to both prevent software aging from changing the results and to minimise the influence of outliers. Preliminary runs indicated severe aging within Sandesha that resulted in exponentially growing transmission times when more than 2,500 messages were sent. These could be traced back to inefficient storage management algorithms used in Sandesha.

Consequently, we tried to avoid these effects. We divided the experiments into runs of 2,000 messages each and restarted both server and client before each run. Since results from these runs are sensitive to random influences, we repeated experiment runs for each oracle ten times per scenario.

In a summarised view (Tables 4.3 and 4.4), no single oracle clearly outperforms all others in all scenarios and with all transports. We will see, however, that the

Table 4.4 Result summary (Mail Transport)

	No Restart	Fixed Int.	Exp. Backoff	Jac./Karn	QEST
No Faults					
avg. ETT	4.63 ± 0.05	4.60 ± 0.05	4.69 ± 0.05	4.64 ± 0.05	4.56 ± 0.05
avg. URC	0.00 ± 0.00	0.05 ± 0.00	0.06 ± 0.00	0.01 ± 0.00	0.02 ± 0.00
10 s Disruption					
avg. ETT	4.72 ± 0.06	4.76 ± 0.05	4.71 ± 0.05	4.69 ± 0.05	4.66 ± 0.05
avg. URC	0.00 ± 0.00	0.06 ± 0.00	0.06 ± 0.00	0.02 ± 0.00	0.00 ± 0.00
60 s Disruption					
avg. ETT	13.79 ± 0.58	11.58 ± 0.38	11.70 ± 0.44	13.37 ± 0.55	11.93 ± 0.46
avg. URC	0.00 ± 0.00	0.61 ± 0.03	0.32 ± 0.01	0.04 ± 0.00	0.12 ± 0.01
2% Packet Loss					
avg. ETT	5.11 ± 0.06	5.16 ± 0.06	5.25 ± 0.06	5.18 ± 0.06	5.15 ± 0.06
avg. URC	0.00 ± 0.00	0.10 ± 0.01	0.12 ± 0.01	0.02 ± 0.00	0.01 ± 0.00
10% Packet Loss					
avg. ETT	9.07 ± 0.17	9.60 ± 0.11	8.97 ± 0.12	8.61 ± 0.14	8.85 ± 0.15
avg. URC	0.00 ± 0.00	0.60 ± 0.01	0.49 ± 0.01	0.03 ± 0.00	0.02 ± 0.00

adaptive oracles perform better under more complex network and system conditions, such as exemplified by the SMTP experiments.

4.5 HTTP Transport

4.5.1 60 s Disruption

The scatter plot in Fig. 4.18 shows average measures for each of the ten runs per oracle in the scenario with a 60 s disruption. ETT is depicted over URC, hence, the closer to the origin they are, the better the oracles perform regarding the tradeoff. We note that all oracles are faster and less fair than the runs without restart, and that there is little variation between runs. The adaptive oracles are fairer and slower than the static ones, and Exponential Backoff performs best in both dimensions.

Both the gain in timeliness through restart and the differences between oracles can be explained by the way TCP handles packet loss. Figure 4.19 shows ETT and URC for messages hit by a 60 s disruption. Without restart, ETT depends entirely on TCP's fault-handling. In this case, we see step-wise constant ETT that start slightly above 90 s and then decrease sharply. This reflects the TCP RTO timeout mechanism. Each transmission attempt required one TCP connection to the server. To set up a connection, TCP engages in a three-way handshake with the server. The initiating party first sends a SYN packet to the destination. If there is no reply, it retransmits the packet. The interval between retransmissions is determined by the RTO, which starts at 3 s and doubles on every retransmission: 6 s, 12 s, ... Waiting times for the connections (and hence for the messages) grow accordingly, for instance $3\,s + 6\,s + \ldots 48\,s = 93\,s$ for messages whose setup phase experiences 5 TCP timeouts, and $3\,s + 6\,s + 12\,s + 24\,s = 45\,s$ if the 4th retransmission is

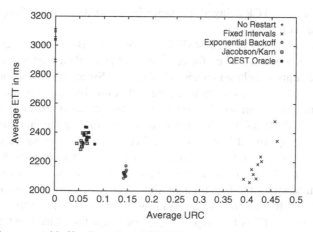

Fig. 4.18 Performance with 60 s disruptions (HTTP Transport)

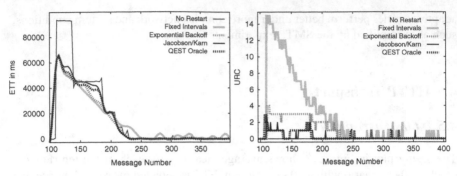

Fig. 4.19 ETT and URC for messages hit by a 60 s disruption (HTTP Transport). ETT curves have been smoothed to improve readability

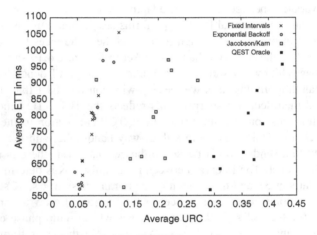

Fig. 4.20 Performance with 2% packet loss (HTTP Transport)

successful. In effect, TCP delayed messages by up to 93 s—much longer than the duration of the fault (60 s).

All oracles yielded ETTs at most slightly above 60 s. By restarting, they initiated new TCP connection setups, i.e., for each restart the TCP entered a new three-way handshake and immediately sent a new SYN packet. Since there were more connection setups, one of them was likely to hit the end of the disruption without engaging the RTO. This connection would then be established faster than one that had to wait for its backed-off RTO to time out to detect the end of the fault and complete its connection setup. Fixed Intervals exemplify this best. A message sent exactly at the beginning of the fault ($t = 60$ s) could be subject to at most $^{60 s}/4 s = 15$ timeouts, the last of which was at $t = 120$ s. Since the fault was over by then, the connection setup for the 16th transmission (15th retransmission) succeeded, and the message could be transmitted almost immediately, with a low transmission time $t_{i16} = r_{i16} - s_{i16}$. ETT is thus $r_{i16} - s_{i1} \approx 60$ s. On the other hand, none of the other transmissions failed. They reached the destination as well, albeit with higher completion times dominated by the TCP (e.g., $t_{i1} \approx 93$ s, see above). Therefore, URC for this message is $16 - 1 = 15$. Exponential Backoff achieved the same ETT in a fairer manner. With restarts at $t = 60 + 4$ s, $60 + 12$ s, $60 + 28$ s, $60 + 60$ s, there were only five transmissions, four of which were unnecessary. Both adaptive oracles' even lower URC is a result of their global timeout. With every elapsed τ_{ij}, be it for a new transmission or one attempted previously, τ grew exponentially, which, due to the number of such events, yielded a very rapid increase. When the first message was again transmitted without a restart, τ_{ij} dropped to about the same value as before the disruption, and then all messages previously held back by the higher timeout were retransmitted at once.

4.5.2 Packet Loss

Without retransmissions, packet loss rates of 2 and 10% led to message loss, i.e., not all messages that were sent reached the destination. This highlights the need for a reliability mechanism on top of HTTP. Although TCP provides reliable connections for HTTP, the HTTP transport can still fail. Again, this can be attributed to the way TCP handles faults. As pointed out above, packet loss in the setup phase delays a connection by at least 3 s. Furthermore, the number of TCP segments exchanged during HTTP transfers is usually small. In consequence, TCP connections that carry HTTP often do not leave the slow-start phase, and thus congestion control prevents fast fault-handling (via duplicate ack detection) from taking effect. (See pp. 303–306 in [86] for details.) With loss rates as high as those studied here, TCP's fault-handling is therefore likely to manifest in connections that are delayed for large amounts of time. Actual implementations, however, cannot wait forever and have to give up eventually. Since HTTP does not retry failed connections, these timeouts transform into message loss.

In regard to oracle performance, we observed different outcomes depending on the packet loss rate. With 2% loss, there were only small differences between oracles (see Table 4.3). Fairness, on the other hand, varied considerably. Both static oracles are much fairer than the adaptive ones. Obviously, more frequent restarts did not help timeliness. If we look at the scatter plot for this scenario (Fig. 4.20), we observe that for all oracles higher URCs tend to correspond to higher ETTs, which is contrary to our notion of a tradeoff between timeliness and fairness. This fact indicates that the costs associated with restart might indeed be high enough to offset its benefits.

Results for the 10% scenario demonstrate that, as the loss rate increases, restart begins to pay off in terms of an ETT improvement. Here, the QEST Oracle as the least fair is also the fastest. Our finding that with 10% packet loss more frequent restarts yield better timeliness is corroborated by the scatter plot (Fig. 4.21), where we note a general trend towards lower ETT with higher URC even within observations for the QEST Oracle.

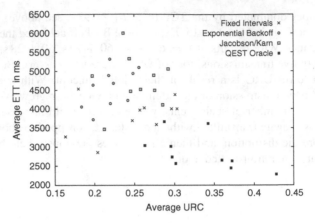

Fig. 4.21 Performance with 10% packet loss (HTTP Transport)

4.5.3 Mail Transport

As with HTTP, the scatter plot for a 60 s disruption when using Mail (Fig. 4.22) shows clearly defined clusters for all oracles. Here, the QEST Oracle performs best, and both static ones are less fair, but not faster. The Jacobson/Karn Oracle exhibits a fairness and timeliness almost identical to runs without restart.

These findings can again be explained by characteristics of the SOAP transport. SMTP servers utilise TCP to transfer mails to their destination. As we laid out in the discussion of HTTP results, TCP connections can be delayed by packet loss. When using the HTTP transport, very long delays lead to SOAP message loss. Unlike the HTTP transport, Mail does employ a reliability mechanism: SMTP servers retry delivery attempts that failed due to non-permanent faults. TCP failures, and failed connection setups in particular, are usually considered transient. The SMTP server

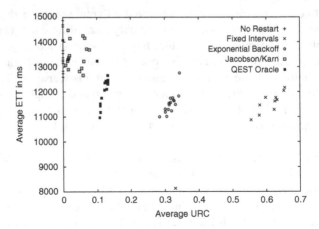

Fig. 4.22 Performance for 60 s disruptions (Mail Transport)

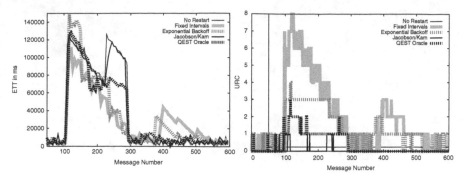

Fig. 4.23 ETT and URC for messages hit by a 60 s disruption (Mail Transport). ETT curves have been smoothed

detects these faults by means of its own timeouts. To the client's SMTP server, the 60 s disruption manifested itself as repeated timeouts when connecting to the destination. The SMTP server then used the queueing system we sketched earlier to retransmit the deferred messages. We can, unfortunately, not endeavour to fully explore the intricate interactions between timeouts and queues in the SMTP server and restarts initiated by the WSRM component. We will thus only point out interesting aspects of the ETT curves shown in Fig. 4.23 and their implications for oracle performance.

What first meets the eye is a pronounced two-peak 'saw-tooth' pattern for transmissions without restart, probably caused by periodic flushing of a queue. Second, both static oracles generally manage to achieve ETTs below these peaks. The apparent gain is especially large with the second peak. Here, both static oracles were about 80 s faster than No Restart. Third, the transport exhibits some sort of memory effect. In all runs, ETT curves for the static oracles sport at least one additional peak with a height of about 40 s and a width of about 120 messages that absorbed previous gains. Fourth, even though the QEST Oracle reduces ETT to roughly 70 s during the second peak, it does not provoke delays in later messages. This fact explains its optimal timeliness. Finally, with the Jacobson/Karn Oracle restarts are almost non-existent (see URC plot in Fig. 4.23), hence the similarity of its results to those without restart.

The recommendation supported by our study is that when using the HTTP transport on top of a simple network, static oracles such as fixed intervals can suffice to maintain a good balance between fairness and timeliness. However, more sophisticated adaptive oracles will be fairer when the system exhibits long periods of packet loss, without being much slower, and faster with continuously high-loss rates. For the Mail transport, which exhibits more intricate timing behaviour, adaptive oracles are a better choice.

In particular, we have seen that the oracle based on the restart model performs in some situations even better than the Jacobson-Karn algorithm.

Chapter 5
Meeting Deadlines Through Restart

Retrying tasks is an obvious thing to do if one suspects a task has failed. However, also if a task has not failed, it may be faster to restart it than to let it continue. Whether restart is indeed faster depends on the completion time distribution of tasks, and on the correlation between the completion times of consecutive tries. As in the previous chapter also in this chapter we assume that the completion times of consecutive tries are independent and identically distributed, an assumption that has been shown to be not unreasonable for Internet applications [128]. Furthermore, we analyse algorithms that are tailored to lognormal distributions, which we (and others) have found to be representative for various Internet applications [86, 128]. Our metric of interest is the probability that a pre-determined deadline is met, and we want to find the restart times that maximise this metric. Note that the metric of meeting deadlines corresponds to points in the completion time distribution, a metric often harder to obtain than moments of completion time which we analysed in the previous chapter. The material in this chapter has been published previously in [178, 164, 177].

We derive two very efficient algorithms to determine the optimal time for restart. The 'equihazard' algorithm finds all restart intervals with equal hazard rates, which corresponds to all local extrema for the probability of making the deadline. We applied the algorithm to lognormal distributed completion times. It turns out that among the equihazard restart intervals in all considered cases equidistant points are optimal. Therefore, a practical engineering approach is to only consider equidistant points, which we do in our second algorithm. The equihazard algorithm finds each local extremum in logarithmic time, the equidistant algorithm takes a constant time to do the same, and finds the globally optimal solution in a few iterations. Hence, these algorithms are excellent candidates for online deployment in potential future adaptive restart implementations.

5.1 A Model for the Probability of Meeting a Deadline Under Restart

To analyse and optimise the time at which to restart a job, we start from a simple model that lends itself to elegant analysis. As in the previous chapter we assume

K. Wolter, *Stochastic Models for Fault Tolerance*,
DOI 10.1007/978-3-642-11257-7_5, © Springer-Verlag Berlin Heidelberg 2010

that the restart of a task terminates the previous attempt and that successive tries are statistically independent and identically distributed. This is for instance the case when we click the reload button in a web browser: the connection with the server is terminated and a new download attempt is tried. In mathematical terms, the problem formulation is as follows. Let the random variable T denote the completion time of a job, with probability distribution $F(t), t \in [0, \infty)$, and let d denote the deadline we set out to meet. Obviously, without restart, the probability that the deadline is met is $F(d)$. Assume τ is the restart time, and the random variable T_τ denotes the completion time when an unbounded number of retries is allowed. That is, a retry takes place periodically, every τ time units, until completion of the job or until the deadline has passed, which ever comes first. We write $f_\tau(t)$ and $F_\tau(t)$ for the density and distribution of T_τ, respectively, and we are interested in the probability $F_\tau(d)$ that the deadline is met.

One can intuitively reason about the completion time distribution with restarts as Bernoulli trials. At each interval between restarts there is a probability $F(\tau)$ that the completion 'succeeds.' Hence, if the deadline d is a multiple of the restart time τ, we can relate the probability of missing the deadline without and with restart through:

$$1 - F_\tau(d) = (1 - F(\tau))^{\frac{d}{\tau}}. \tag{5.1}$$

If the restart intervals are not identical we denote their length by $\tau_1, \ldots, \tau_n$, assuming n intervals. If furthermore d is not exactly reached by the last interval the remaining time after the last interval until the deadline is $d - \sum_{i=1}^{n} \tau_i$. If we furthermore introduce a penalty, or cost c associated with restart the probability of missing the deadline without and with restart relate similarly to the probability distribution function defined in (4.3):

$$1 - F_\tau(d) = \begin{cases} \prod_{i=1}^{k}(1 - F(\tau_i)) \cdot (1 - F(d - \sum_{i=1}^{k}(\tau_i + c))) & \text{if } \sum_{i=1}^{k}(\tau_i + c) \le d < \sum_{i=1}^{k+1} \tau_i + kc \\ \prod_{i=1}^{k+1}(1 - F(\tau_i)) & \text{if } \sum_{i=1}^{k+1} \tau_i + kc \le t < \sum_{i=1}^{k+1}(\tau_i + c). \end{cases} \tag{5.2}$$

For the sake of an easier treatment we will in the following assume that $c = 0$ and $\tau_1 = \ldots = \tau_n = \tau$, and that the deadline is an integer multiple of the restart time $d = n\tau$.

For a single retry during the finite interval $[0, d)$, when the retry is at time τ, $\tau < d$, then the probability of completion before d is:

$$F_\tau(d) = 1 - (1 - F(\tau))(1 - F(d - \tau)). \tag{5.3}$$

By equating the derivative with respect to τ to zero, we obtain for the extrema of $F_\tau(d)$ that:

$$\frac{f(\tau)}{1 - F(\tau)} = \frac{f(d - \tau)}{1 - F(d - \tau)}. \tag{5.4}$$

The function

$$h(t) = \frac{f(t)}{1 - F(t)}$$

is known as the hazard rate, and is key throughout our analysis and algorithms. The above result shows that minima and maxima for the probability that a deadline is met with restarts are found at *equihazard* restart intervals. Moreover, the *equidistant* restart intervals $\tau = \frac{d}{2}$ are a special case of equihazard intervals, and form thus also a local extremum.

For multiple retries before the deadline similar mathematics can be applied. This time we take derivatives with respect to each restart interval $\tau_i, i = 1, \ldots, N$. (Note, the restarts take place at times $\tau_1, \tau_1 + \tau_2, \ldots, \sum_{n=1}^{N} \tau_n$, and we assume without loss of generality that $\sum_{n=1}^{N} \tau_n = d$.) Then we obtain that an optimum with respect to all retry intervals $\tau_1, \ldots, \tau_N$ is found when:

$$\frac{f(\tau_1)}{1 - F(\tau_1)} = \frac{f(\tau_2)}{1 - F(\tau_2)} = \ldots = \frac{f(\tau_N)}{1 - F(\tau_N)}. \tag{5.5}$$

Again, the extrema are at equihazard intervals, with as special case the equidistant restart intervals $\tau_n = \frac{d}{N}$.

5.2 Algorithms for Optimal Restart Times

Very often, completion times for Internet tasks have a distribution function that can be closely fit by a lognormal distribution [86, 128]. Since the Internet is one of our anticipated application fields we chose in this section the lognormal distribution with parameters we fit to the data in [128]. The density function and the hazard rate of a lognormal distribution are shown in the Appendix in Sect. B.2.6.

The lognormal shape of the hazard function can be exploited by optimisation algorithms, since it has at most two points with the same hazard function value. This allows us to quickly identify all potential solutions of the optimisation problem. The following algorithm finds the two restart interval lengths τ_a and τ_b for which holds:

$$h(\tau_a) = h(\tau_b), \tag{5.6}$$
$$n_a \tau_a + n_b \tau_b = d, \tag{5.7}$$

where n_a and n_b denote the number of intervals of each length. The parameters n_a and n_b are input to the algorithm, and to find the optimal restart strategy, one needs to call the algorithm for all relevant combinations of n_a and n_b, and then select from all the equihazard solutions the one that optimises the probability of meeting the deadline.

Algorithm 5 (Equihazard Restart Intervals)

```
Input n_a and n_b;
top = d/n_b; bottom = d/(n_a + n_b);
τ_b = top; τ_a = (d-n_b τ_b)/n_a;
Repeat {
   top = (top+bottom)/2;
   τ_b = top;
   τ_a = (d-n_b τ_b)/n_a; (so interval lengths sum to d)
   If( SignChanged(h(τ_b) - h(τ_a)) ) {
      bottom = top;
      top = PreviousValue(top);
   }
}
Until (top-bottom ≈ 0)
```

To explain the working of Algorithm 5, first note that one solution to (5.7) is the equidistant restart strategy $\tau_a = \tau_b = \frac{d}{N}$. The algorithm will end up with that solution, unless there exists a second solution. For this solution, it cannot be that τ_a and τ_b are both smaller or both larger than $\frac{d}{N}$, since then the intervals would not sum to d. Therefore, we can choose $\tau_b > \frac{d}{N}$ and $\tau_a < \frac{d}{N}$. Furthermore, it also must hold that $\tau_b \le \frac{d}{n_b}$. The algorithm utilises these facts to initialise an interval between bottom and top in which τ_b lies, and then breaks the interval in two at every iteration, until top ≈ bottom. At every iteration, it sets τ_b to the guess top and computes the belonging $\tau_a = \frac{d-n_b \tau_b}{n_a}$. It then tests if the sign of $h(\tau_b) - h(\tau_a)$ changes, to decide if τ_b lies in the upper or lower half. This test works correctly thanks to the particular shape of the lognormal hazard function. Note that since the algorithm divides the considered interval in two in every iteration, it takes logarithmic time to find the optimum for every pair n_a, n_b for which the algorithm is run.

We applied Algorithm 5 to the lognormal distribution with parameters $\mu = -2.3$ and $\sigma = 0.97$, and deadline $d = 0.7$. The parameters fit data collected in [128], but are otherwise arbitrary. Figure 5.1 shows typical behaviour if one considers a single restart. The equidistant restart (at $\tau = 0.35$) is optimal, while the other equihazard points turn out to be minima ($\tau = 0.013$ or $\tau = 0.687$). The improvement in probability of making the deadline is from 0.977 to 0.990. Table 5.1 shows results for increasing number of restarts, displaying all sets of equihazard intervals that are extrema. We see from the table that for this example equidistant hazard rates always outperform the other equihazard points, and that the optimum is for three equidistant restarts (and thus four intervals). We also see from the table that if we restart too

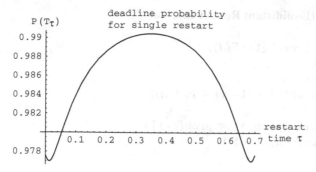

Fig. 5.1 Probability of meeting deadline for one restart ($d = 0.7$, $\mu = -2.3$, $\sigma = 0.97$)

early (using the minima in Fig. 5.1) the probability of meeting the deadline is even less than without any restart.

It turns out that equidistant restarts are optimal in all experiments with lognormal distributions. Although we can construct examples in which for instance two non-equidistant points outperform equidistant points, for the lognormal distribution this only seems to be possible if no restart performs even better. Unfortunately, we have no proof for this phenomenon, but it gives us ground to use an algorithm that limits its search for optima to equidistant points, which can be done even faster than Algorithm 5 for equihazard points. In the following algorithm we increase the number of equidistant restart points (starting from 0), consider the probability of making the deadline for that number of restarts and stop as soon as we see no more improvement. This is a very advantageous stopping criterion since one needs not to set an arbitrary maximum on the number of restart points. We do not give the derivation of the correctness of this stopping criterion here, but instead close the discussion with the algorithm.

Table 5.1 Equihazard restart intervals and associated probability of meeting the deadline ($d = 0.7$, $\mu = -2.3$, $\sigma = 0.97$)

# restarts	Equihazard intervals	$P(T_{\{\tau\}} < d)$
0	—	0.978
1	0.35, 0.35	0.990
1	0.013, 0.687	0.977
2	0.23, 0.23, 0.23	0.993
2	0.019, 0.34, 0.34	0.990
2	0.013, 0.013, 0.674	0.976
3	0.175, 0.175, 0.175, 0.175	0.99374
3	0.024, 0.225, 0.225, 0.225	0.993
3	0.019, 0.019, 0.331, 0.331	0.989
3	0.013, 0.013, 0.013, 0.660	0.976
4	0.14, 0.14, 0.14, 0.14, 0.14	0.99366
⋮	⋮	⋮

Algorithm 6 (Equidistant Restart Intervals)

```
n=1; prob[1]=F(d);
Do{
  n++;
  prob[n] = 1 - (1 - F(d/n))^n;
}
Until (prob[n] < prob[n-1])
Return(d/(n - 1))
```

Where other mechanisms like early fault detection add an enormous computation overhead to increase availability [133] we are able to increase the probability of completion before a deadline with a very easy mechanism by a factor of two ('half a nine'). Further study is needed not only to see how the algorithms generalise to other completion time distributions but also how the impact of restarts depends on the remoteness of the deadline.

5.3 An Engineering Rule to Approximate the Optimal Restart Time

The algorithms for finding the optimal restart time to maximise the probability of meeting a deadline work well for theoretical distributions of a certain shape. However, for empirical data we might want a more rough approximation of (5.1) to be able to implement online methods for finding those optimal restart times, as we will show in the following section. And indeed, an approximation of (5.1) helps us to find a simple rule which we use in an online fashion.

As in Sect. 5.1 we again need some definitions: Let the random variable T denote the completion time of a job, with probability distribution $F(t)$, $t \in [0, \infty)$. Assume τ is a restart time, and the random variable T_τ denotes the completion time when an unbounded number of retries is allowed. We write $f_\tau(t)$ and $F_\tau(t)$ for the density and distribution of T_τ. We also use the hazard rate

$$h(t) = \frac{f(t)}{1 - F(t)}. \tag{5.8}$$

New in this section is the use of the cumulative hazard

$$H(t) = \int_{s=0}^{t} h(s)ds,$$

which is closely related to the distribution function in that

$$1 - F(t) = e^{-H(t)} \tag{5.9}$$

or

$$H(t) = -\log(1 - F(t)). \tag{5.10}$$

Restart at time τ is beneficial only if the probability $F_\tau(d)$ of making the deadline d under restart is greater than the probability of making the deadline without restart, i.e.

$$F_\tau(d) > F(d). \tag{5.11}$$

We again start from (5.1) which for completeness we repeat here

$$1 - F_\tau(t) = (1 - F(\tau))^{\frac{d}{\tau}}. \tag{5.12}$$

If there would be a restart time τ that maximises the completion probability $F_\tau(t)$ for all values of t, this would be the ideal restart time and be 'stochastically' optimal. However, except for pathological cases, such a restart time does not exist. Equation (5.12) is correct only for values of d and τ such that d is an integer multiple of τ. But if we ignore this fact, or simply accept (5.12) as an approximation, we can find the optimal restart time in a straightforward way. Surprisingly, it turns out that the approximation gives us a restart time independent of the deadline d, which is optimal in the limit $d \to \infty$. That is, it optimises the tail of the completion time distribution under restarts, and is therefore beneficial for many other metrics as well, such as higher moments of the completion time.

Theorem 5.1 *The following statements about our approximation τ^* are equivalent.*

1. τ^* is an extremum (in τ) of

$$(1 - F(\tau))^{\frac{d}{\tau}} \tag{5.13}$$

for any *deadline d;*
2. τ^* is the point where

$$\tau^* \cdot h(\tau^*) = -\log(1 - F(\tau^*)); \tag{5.14}$$

3. τ^* is a point where

$$\tau^* \cdot h(\tau^*) = H(\tau^*); \tag{5.15}$$

4. τ^* is an extremum of

$$\frac{-\log(1 - F(\tau))}{\tau}; \tag{5.16}$$

5. τ^* *is an extremum of*

$$(1 - F(\tau))^{\frac{1}{\tau}}. \tag{5.17}$$

Proof We use

$$\frac{d}{dx}(g(x))^x = (g(x))^x \left(\frac{x \frac{d}{dx} g(x)}{g(x)} + \log(g(x)) \right).$$

If the first item is true, then τ^* is an extremum when the derivative of $(1 - F(\tau))^{\frac{d}{\tau}}$ equates to 0:

$$\frac{d}{d\tau}(1 - F(\tau))^{\frac{d}{\tau}} = (1 - F(\tau))^{\frac{d}{\tau}} \left(\frac{f(\tau)\tau}{1 - F(\tau)} + \log(1 - F(\tau)) \right) = 0.$$

Irrespective of the value of d, Statement 2 then follows immediately:

$$\frac{f(\tau^*)}{1 - F(\tau^*)} = \frac{-\log(1 - F(\tau^*))}{\tau^*},$$

which can be rewritten using (5.8) into

$$\tau^* \cdot h(\tau^*) = -\log(1 - F(\tau^*)),$$

and thus Statement 2 holds if and only if Statement 1 holds.

The equivalence between Statements 2 and 3 follows using the relation (5.10). Statement 4 follows from taking the derivative of $\frac{-\log(1 - F(\tau))}{\tau}$ and equate it to 0:

$$\frac{d}{d\tau} \frac{-\log(1 - F(\tau))}{\tau} = -\frac{1}{1 - F(\tau)}(-f(\tau))\tau - \log(1 - F(\tau)) = 0$$

$$\Longleftrightarrow \qquad \frac{f(\tau)}{1 - F(\tau)} = \frac{\log(1 - F(\tau))}{\tau}.$$

So, Statement 4 holds if and only if statement 1 holds.

Statement 5 then follows from taking exponential power of the expression in Statement 4

$$\exp\left(-\log(1 - F(\tau)) \cdot \frac{1}{\tau} \right) = (1 - F(\tau))^{\frac{1}{\tau}}.$$

Note that, alternatively, Statements 4 and 5 can be obtained from properly manipulating the result in Statement 1 for $d = 1$. □

The heuristics behind this approximation is that (1) the optimal restart times correspond to equihazard points; (2) equidistant restart times are equihazard points and often (albeit not always) optimal. The mathematical trick then is to relax the restriction that the number of restarts must be integer valued. In doing so, one obtains a continuous function, for which one can take derivatives and get relations for its extremes. If one carries this out, it turns out that the optimal restart time is independent of the time d one wants to optimise $(1 - F(\tau))^{\frac{d}{\tau}}$ for. Moreover, if $\frac{d}{\tau^*}$ takes an integer value, restart time τ^* is an equidistant restart strategy, and thus a local extremum. From this reasoning it also follows that the rule gets closer to the optimum for the tail of $F_\tau(t)$, since then $\frac{t}{\tau^*}$ is close to an integer value. This results in the following claim, which is stated without proof:

Theorem 5.2 *For $d \to \infty$, the approximation error in the restart interval lengths converges to zero:*

$$d - n^* \cdot \tau^* \to 0. \qquad (5.18)$$

where n^* is the maximum number of intervals of length τ^* that can be accommodated in the interval $[0, d]$

$$n^* = \left\lfloor \frac{d}{\tau^*} \right\rfloor.$$

Item 3 of Theorem 5.1 can be interpreted in the following way: the surface under the hazard rate curve up to point τ^* equals the rectangle defined by x- and y-value of $h(\tau^*)$ as illustrated in Fig. 5.2. We will refer to (5.15) as the *rectangle equals surface rule*. This very illustrative and simple rule is used later in a pragmatic algorithm for an empirical hazard rate to find an empirical optimal restart time that maximises the probability of completion, the probability of making an infinite deadline.

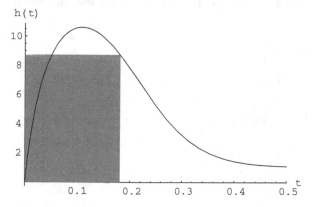

Fig. 5.2 Illustration of the surface = rectangle rule, with optimal restart time 0.18 (hyper/hypo-exponential distribution with parameters as indicated in Appendix B.2.4)b

The quality of the approximation of the equivalent formulas in Theorem 5.1 is evaluated by formulating an engineering rule based on (5.16) (Item 4 of Theorem 5.1) and using this rule in experiments. The rule is defined as follows.

Algorithm 7 *Set the restart time at τ^* with τ^* the optimum of $\frac{-log(1-F(d))}{d}$.*

It should be noted that if the hazard rate is monotonously increasing, no value of τ exists that satisfies Theorem 5.1. In this case restart will not help increasing the probability of completion. Whereas if the hazard rate is monotonously decreasing the rectangle equals surface rule holds only for $\tau = 0$, which means immediate restarts. Most other distributions will have a single maximum in their hazard rates, after which the hazard rate then decreases. Although, of course, distributions can be constructed that have several local maxima also in their hazard rates. In both latter cases after some point a value of τ exists, such that (5.15) holds. Only then restart can be applied successfully.

Figure 5.3 shows a variety of optimal restart times. The example is the lognormal distribution we already used in the previous chapter, with parameters $\mu = -2.3$, and $\sigma = 0.97$. The straight lines are the approximation, and the optimal restart times to minimise the first and second moment of the completion time. The sea-saw line is the optimal restart time for points on the distribution. That is, the x-axis gives the point t on the distribution, the y-axis the restart time τ that maximises $F_\tau(t)$. We see that the approximation gets closer to optimal as t increases; in other words, the approximation works best for the tail of the completion time distribution. This fact suggests that the approximation may be better for higher moments of completion time, since these are more sensitive to the tail of the distribution. This seems indeed to be the case, since the restart time that minimises the first moment is farther off our approximation than the restart time that minimises the second moment.

Figure 5.4 shows the completion time distribution for the various restart regimes. The solid curve gives $F_\tau(t)$ with t on the x-axis and τ equal to our approximation only for the interval $t \in [0.0, 0.5]$. The dotted curve uses the restart time at point

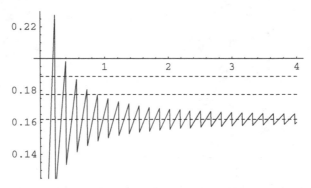

Fig. 5.3 Optimal restart times: approximation (*lowest*), optimal for first moment (*highest*), optimal for second moment (*middle*), optimal for points in distribution (*sea-saw*)

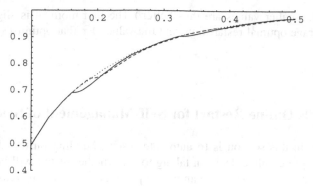

Fig. 5.4 $F_\tau(t)$ using optimal restart time at t (*dotted*), approximation (*solid line*), and using the optimal restart time for first moment (*dashed*)

t that optimises $F_\tau(t)$. This is the theoretical optimum. The dashed curve uses the restart time that minimises the first moment of completion time. We see that both the approximation and the optimum restart time for the first moment are very close to the completion time distribution using the theoretically optimal τ. Figure 5.5 shows on a logscale the difference between the completion time distributions $F_\tau(t)$ over a larger time horizon when computing τ using the approximation and the optimal completion time, as well as the difference between the completion time with restart optimised for the first moment and the theoretical optimum. We see that the approximation is exact at the spikes, which appear with distance equal to τ^*. We also see that if t increases, the optimal restart time for the first moment never reaches the theoretical optimum, and in fact slowly diverges from the approximation.

Table 5.2 shows how three different restart times perform with respect to moments and quantiles of the completion time distribution under restarts. One can see that the differences are minor, our approximation performing best for the 90% and higher quantiles. (The value in the table under quantiles is the point t at which $F_\tau(t) = 0.9$

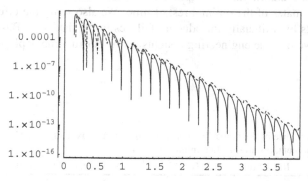

Fig. 5.5 Logarithm of differences in completion time distribution $F_\tau(t)$ when using τ as computed for optimal restart times and approximation (*solid line*), using τ computed for optimal restart time and optimised for first moment (*dashed*)

(0.99, 0.999), smaller values are thus better.) The first moment is slightly lower using the average optimal restart time and the value of τ that optimises the second moment.

5.4 Towards Online Restart for Self-Management of Systems

The objective in this section is to automate restart, building on the above work. We want to explore online decision taking to see whether restart will be beneficial and when to apply it. We simulate an online procedure by using increasingly more data from measurements taken earlier [128], but the applied methods can easily be included in a software module like the proxy server in [128] to be executed in real-time.

In Sect. 3.2 we have already seen that the shape of the hazard rate of a probability distribution indicates whether restart is beneficial. For empirical data the correct theoretical distribution is unknown and the hazard rate therefore needs to be estimated based on observations. Estimating the hazard rate is not a straightforward task, since it needs numerical computation of the derivative of the cumulative hazard rate. We derive and implement our simple rule based on the hazard rate that allows us to find the optimal restart time to maximise the probability of making a deadline. The rule approximates the optimal restart time independent of the exact value of the deadline, and is asymptotically exact (when the deadline increases). Moreover, the rule is very simple, making it a likely candidate for runtime deployment. One has to bear in mind that not in all cases the optimal restart time does exist. Restart is applicable to a system if (and only if) the rule finds an optimal restart time. So, our simple rule actually serves a two-fold purpose: it enables us to decide whether restart will be beneficial in the given situation, and if so, it provides us with the optimal restart time.

We apply the rule to data sets we collected for HTTP, thus mimicking the online execution of the algorithm. We explore how much data is required to arrive at reasonable estimates of the optimal restart time. We also study the effect of failed HTTP requests by artificially introducing failures in the data sets. Based on these explorations we provide engineering insights useful for runtime deployment of our algorithm.

Table 5.2 Performance of three possible restart times: our approximation, restart time that minimises second moment, restart time that minimises first moment

	Restart time	$E[T_\tau]$	$E[T_\tau^2]$	90% quant.	99% quant.	99.9% quant.
Approximation	0.162	0.14600	0.0399	0.317	0.635	0.952
2nd Moment opt.	0.177	0.14573	0.0398	0.318	0.639	0.962
Average optimal	0.189	0.14568	0.0399	0.319	0.645	0.970

5.4.1 Estimating the Hazard Rate

It follows from the *surface equals rectangle* rule (5.15) that an estimate $\hat{h}(t)$ of the hazard rate curve is needed to determine the optimal restart time. We will in this section provide the main steps of how to estimate the hazard rate and implement the rule (5.15) in an algorithm. Some details are shifted to Appendix B.2.6. We use the theory on survival analysis in [80]. The hazard rate $h(t)$ cannot be estimated directly from a given data set. Instead, first the cumulative hazard rate $H(t)$ is estimated and then the hazard rate itself is computed as a numerical derivative.

Let us consider a sample of n individuals, that is n completions in our study. We sample the completion times and if we order them, we obtain a data set of D distinct times $t_1 \le t_2 \le \ldots \le t_D$ where at time t_i there are d_i events, that is d_i completions take time t_i. The random variable Y_i counts the number of jobs that need more or equal to t_i time units to complete. We can write Y_i as

$$Y_i = n - \sum_{j=1}^{i-1} d_j$$

All observations that have not completed at the end of the regarded time period, usually time t_D, are called *right censored*. There are $Y_n - d_n$ right censored observations. The experimental data we use falls in that category, since Internet transactions commonly use TCP, which aborts (censors) transactions if they do not succeed within a given time.

The hazard rate estimator $\hat{h}(t)$ is the derivative of the cumulative hazard rate estimator $\hat{H}(t)$, which is defined in Appendix C.1. It is estimated as the slope of the cumulative hazard rate. Better estimates are obtained when using a kernel function to smooth the numerical derivative of the cumulative hazard rate. The smoothing is done over a window of size $2b$. A bad estimate of the hazard rate will yield a bad estimate of the optimal restart time and the optimised metric is very sensitive to whether the restart time is chosen too short. Therefore obtaining a good estimate of the hazard rate is important.

Let the magnitude of the jumps in $\hat{H}(t)$ and in the estimator of its variance $\hat{V}[\hat{H}(t)]$ at the jump instants t_i be $\Delta \hat{H}(t_i) = \hat{H}(t_i) - \hat{H}(t_{i-1})$ and $\Delta \hat{V}[\hat{H}(t_i)] = \hat{V}[\hat{H}(t_i)] - \hat{V}[\hat{H}(t_{i-1})]$. Note that $\Delta \hat{H}(t_i)$ is a crude estimator for $\hat{h}(t_i)$.

The kernel-smoothed hazard rate estimator is defined separately for the first and last points, for which $t - b < 0$ or $t + b > t_D$. For inner points with $b \le t \le t_D - b$ the kernel-smoothed estimator of $h(t)$ is given by

$$\hat{h}(t) = b^{-1} \sum_{i=1}^{D} K\left(\frac{t - t_i}{b}\right) \Delta \hat{H}(t_i). \tag{5.19}$$

The variance of $\hat{h}(t)$ is needed for the confidence interval and is estimated as

$$\sigma^2[\hat{h}(t)] = b^{-2} \sum_{i=1}^{D} K\left(\frac{t - t_i}{b}\right)^2 \Delta\hat{V}[\hat{H}(t_i)].vspace * 12pt \qquad (5.20)$$

The function $K(.)$ is the Epanechnikov kernel defined in Appendix C.2.

A $(1 - \alpha) \cdot 100\%$ point wise confidence interval around $\hat{h}(t)$ is constructed as

$$\left[\hat{h}(t)\exp\left[-\frac{z_{1-\alpha/2}\sigma(\hat{h}(t))}{\hat{h}(t)}\right], \hat{h}(t)\exp\left[\frac{z_{1-\alpha/2}\sigma(\hat{h}(t))}{\hat{h}(t)}\right]\right]. \qquad (5.21)$$

where $z_{1-\alpha/2}$ is the $(1 - \alpha/2)$ quantile of the standard normal distribution.

The choice of the right bandwidth b is a delicate matter, but is important since the shape of the hazard rate curve greatly depends on the chosen bandwidth (see Fig. 5.7) and hence a badly chosen bandwidth will have a serious effect on the optimal restart time. One way to pick a good bandwidth is to use a cross-validation technique of determining the bandwidth that minimises some measure of how well the estimator performs. One such measure is the *mean integrated squared error* (MISE) of $\hat{h}$ over the range τ_{min} to τ_{max}. The mean integrated squared error can be found in Appendix C. To find the value of b which minimises the MISE we find b which minimises the function

$$g(b) = \sum_{i=1}^{M-1}\left(\frac{t_{i+1} - t_i}{2}\right)(\hat{h}^2(t_i) + \hat{h}^2(t_{i+1})) - 2b^{-1}\sum_{i \neq j} K\left(\frac{t_i - t_j}{b}\right)\Delta\hat{H}(t_i)\Delta\hat{H}(t_j). \qquad (5.22)$$

Then $g(b)$ is evaluated for different values of b. Each evaluation of $g(b)$ requires the computation of the estimator of the hazard rate. The optimal bandwidth can be determined only in a trial-and-error procedure. We found in our experiments that the optimal bandwidth is related with the size of the data set and the variance of the data. We use the standard deviation to determine a starting value and then do a simple step-wise increase of the bandwidth until $g(b)$ takes on its minimal value. In case the hazard rate is increasing in the first steps, we decrease b and start again, since then we are obviously beyond the minimum already. In our experiments and in the literature we always found a global minimum, never any local minima. Advanced hill-climbing algorithms can be applied to find the minimum more quickly and more accurately than we do here.

Once the best estimate of the hazard rate is found we need to determine the point i^* that satisfies the *rectangle equals surface* rule (5.15).

The following simple algorithm determines the optimal restart time τ^* by testing all observed points $t_i, i = 1, \ldots, n$ as potential candidates.

Algorithm 8 (Optimal restart time)

```
Input ĥ, Ĥ and t;
i = 1; #(t = t₁, ..., tₙ)
While((i < n) and (tᵢ · ĥ(tᵢ) > Ĥ(tᵢ)) ) {
        i++;
}
return tᵢ;
```

This algorithm returns in the positive case the smallest observed value that is greater than the estimated optimal restart time τ^*.

In many cases, however, the studied data set does not contain observations large enough to be equal or greater than the optimal restart time. Then we extrapolate the estimated hazard rate to find the point where the rectangle equals the surface under the curve. Assuming we have a data set of n observations $t_i, i = 1, \ldots, n$, at first the slope of the estimated hazard rate at the end of the curve is determined as the difference quotient

$$\text{slope} = \frac{\hat{h}(t_n) - \hat{h}(t_{n-1})}{t_n - t_{n-1}}. \tag{5.23}$$

Then $t_\tau = t_n + \Delta t$ is determined such that for t_τ Eq. (5.15) holds.

$$(t_n + \Delta t) \cdot (\hat{h}(t_n) + \text{slope} \cdot \Delta t) = \hat{H}(t_n) \cdot \text{slope} \cdot \Delta t \cdot t_n$$

$$\iff \quad \Delta t = \frac{\hat{H}(t_n) - t \cdot \hat{h}(t_n)}{\hat{h}(t_n) - 2\text{slope}t_n - \hat{H}(t_n) - \text{slope}}. \tag{5.24}$$

The computational complexity of the algorithm depends in first place on the number of iterations needed to find the optimal bandwidth for the hazard rate estimator. In our experiments we used a heuristic based on the standard deviation of the data set that gave us the optimal bandwidth often in less than 5 iterations, but sometimes took up to 20 iterations.

The second important parameter is the number of observations considered. Each iteration on the bandwidth requires the computation of the estimated hazard rate, which in turn needs traversing all observations and uses for each point a window of size $2b$. Complexity of the hazard rate estimator is therefore at most $O(n^2)$. Improving on the heuristic for the bandwidth, so that in all cases only few iterations are needed is certainly worth while.

5.4.2 Experiments

We have implemented the algorithm to estimate the hazard rate and determine the optimal restart time as defined by (5.15) in Theorem 5.1. The implementation is done in Mathematica and has been applied to the HTTP connection setup data studied in [128]. This data in fact consists of the time needed for TCP's three-way handshake to set up a connection between two hosts.

In our experiments we investigate various issues. One is the uncertainty introduced by small sample sizes. The available data sets consist of approximately one thousand observations for each URL, that is thousand connection setup times to the same Internet address. We use these data sets and take subsets of first one hundred then two hundred observations etc. as indicated in the caption of the figure and in the table. We do not use data of different URLs in one experiment since we found that very often different URLs have different distributions or at least distribution parameters. Furthermore, the application we have in mind is web transactions between two hosts.

The data we study is Data Set '28' consisting of roughly 1,000 connection setup times to http://nuevamayoria.com, measured in seconds. This data set shows characteristics such as a lower bound on all observation and a pattern of variation which we found in many other data sets as well, even though usually not with the same parameters. The chosen data set is therefore to be seen as one typical representative of a large number of potential candidates. The considered connection setup times are shown in Fig. 5.6. The largest observation in this data set is 0.399678 s.

For each of the mentioned subsamples the optimal smoothing factor, or bandwidth, is computed by evaluating (5.22) several times, finding the minimum in a simple search. Figure 5.7 shows estimates of the hazard rate for different values of the bandwidth. Parameter $b1$ is too large, whereas $b2$ is too small, $b3$ is the one that minimises the error and is therefore the optimal bandwidth. One can see that too large a bandwidth leads to an extremely smooth curve, whereas too small a bandwidth produces over-emphasised peaks. From the figure one might conclude that rather too large a bandwidth should be chosen than one that is too small, but more experiments are needed for a statement of this kind. Using the optimal bandwidth,

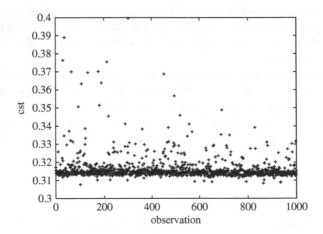

Fig. 5.6 Data Set No. 28; connection setup times (in seconds)

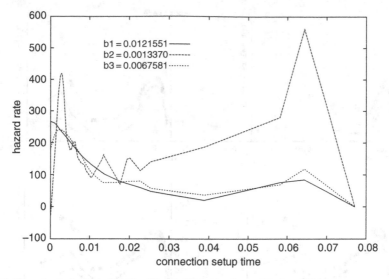

Fig. 5.7 Hazard rate for Data Set No. 28 and different values of the bandwidth b

the hazard rate and its 95% confidence interval are estimated according to (5.19) and (5.21). Finally, for each estimated hazard rate the optimal restart time τ^* is computed using Algorithm 8. In some cases, the algorithm finds the optimal restart time, since the data set includes still an observation greater than the optimal restart time. If the data set has no observation large enough to be greater than the optimal restart time, we extrapolate according to (5.24). The optimal restart times are drawn as vertical bars in the plots in Figs. 5.8 and 5.9.

Note that in Fig. 5.8 although it looks like all optimal restart times are extrapolated in fact none of them is. The extrapolated optimal restart times are indicated by an asterisk in Table 5.3.

The hazard rate curve has no value at the point of the largest observation, since for the numerical derivation always two data points are needed. Furthermore, because of the limited amount of data in the tail, it is not surprising that the confidence interval at the last observations grows rapidly.

Table 5.3 shows some characteristics obtained in the program runs for Data Set 28. Each block of the table belongs to a subset of size n with corresponding standard deviation. The standard deviation changes as more observations come into consideration. For each subsample three different cases are studied. In the first one only the n observations are used and the failure probability equals either zero, or the relative fraction of observations that are greater than 3.0. This threshold is the first retransmission timeout of TCP and hence observations greater 3.0 are (somewhat arbitrarily) censored and retried. We treat them as censored observations and all censored observations contribute to the failure probability. Data Set '28' does not have any such censored observations, but many other data sets do. The second group consists of the n observations plus $2n$ censored ones and has therefore failure prob-

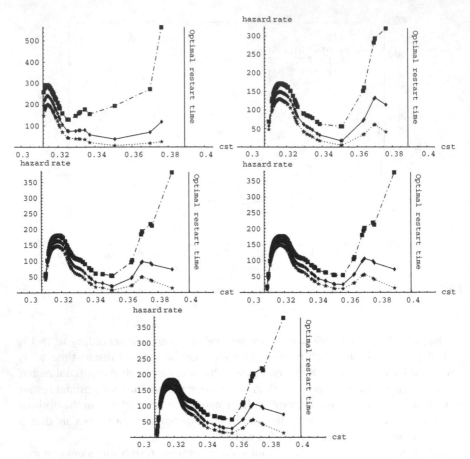

Fig. 5.8 Estimated hazard rates and confidence intervals for the estimates for increasing sample size (*top* row $n = 100$ and $n = 200$, *middle* row $n = 400$ and $n = 600$, *bottom* row $n = 800$) and failure probability 0.0

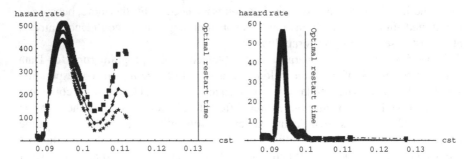

Fig. 5.9 Estimated hazard rates and confidence intervals for sample size $n = 1,000$, failure probability 0.0 (*left*) and 0.8 (*right*)

Table 5.3 Optimal restart time (τ^*) and optimal bandwidth (bw) for different subsample sizes of Data Set 28 and different failure probabilities

$n = 100$, StdDev = 0.0121551			$n = 200$, StdDev = 0.0117341		
Failure prob.	bw	τ^*	Failure prob.	bw	τ^*
0.0	0.006758	0.389027	0.0	0.011557	0.389027
0.666667	0.001779	0.597251*	0.666667	0.001398	0.674306*
0.8	0.001779	0.554513*	0.8	0.001271	0.638993*
$n = 300$, StdDev = 0.0106746			$n = 400$, StdDev = 0.010383		
Failure prob.	bw	τ^*	Failure prob.	bw	τ^*
0.0	0.011742	0.389027	0.0	0.010226	0.399678
0.666667	0.001272	0.333271	0.666667	0.001124	0.333271
0.8	0.001156	0.333271	0.8	0.001124	0.333271
$n = 500$, StdDev = 0.00997916			$n = 600$, StdDev = 0.00941125		
Failure prob.	bw	τ^*	Failure prob.	bw	τ^*
0.0	0.010977	0.399678	0.0	0.010352	0.399678
0.666667	0.001081	0.333271	0.666667	0.001138	0.333271
0.8	0.001081	0.333271	0.8	0.001019	0.333271
$n = 700$, StdDev = 0.00895504			$n = 800$, StdDev = 0.00851243		
Failure prob.	bw	τ^*	Failure prob.	bw	τ^*
0.0	0.009850	0.309209	0.0	0.0103	0.399678
0.66667	0.000970	0.333271	0.66667	0.000922	0.332014
0.8	0.000970	0.333271	0.8	0.000922	0.332014
$n = 900$, StdDev = 0.00816283			$n = 1000$, StdDev = 0.00784583		
Failure prob.	bw	τ^*	Failure prob.	bw	τ^*
0.0	0.009877	0.308456	0.0	0.009493	0.308456
0.6667	0.000884	0.332014	0.6667	0.000949	0.333271
0.8	0.000884	0.332014	0.8	0.000850	0.332014

ability 2/3, or a little higher if there are additional censored observations present in the data set. Analogously, the third group has $n + 4n$ observations and a failure probability of $4n/5n = 0.8$ (or more if there are censored observations in the data set).

When looking at the results for failure probability zero, also plotted in Fig. 5.8 for $n = 100, 200, 400, 600, 800$ we see that the small data sets lead to an overestimated optimal restart time (if we assume that the full 1,000 observations give us a *correct* estimate), and the 'correct' value is overestimated by less than 5%.

Such high failure probability can be interpreted as an interruption of the connection. It should be noted that a failure probability of 0.1 or less does not show in the results at all. Looking at the results for the different sample sizes in the group with high failure probability, we also find that with the small samples the optimal restart time is overestimated.

The impact of the failure probability within a group of fixed sample size has been investigated as well. The failure probability is increased by subsequently adding more failed (and hence censored) observations and then estimates for the hazard rate and optimal restart time are computed. The failed attempts of course increase the sample size. We notice (as can be seen in Table 5.3) that the bandwidth used for estimating the hazard rate decreases for increasing failure rate, while the sample standard deviation is computed only from non-failed observations and hence does not change with changing failure probability. We found in [163] that for theoretical distributions the optimal restart time decreases with increasing failure probability.

Typically, our experiments agree with this property, which, however, is not true for some subsets of Data Set '28'.

An additional purpose of the experiments was to find out whether we can relate the optimal bandwidth to any characteristic of the data set. In the literature no strategy is pointed out that helps in finding the optimal bandwidth quickly. In our implementation we set the standard deviation as a starting value for the search. If we have no censored observations (failure probability zero) we always find the optimal bandwidth within less than five iterations. If the data set has many censored observations the optimal bandwidth roughly by factor 5 and we need more iterations to find that value, since our heuristic has a starting value far too large in that case.

Figure 5.9 compares two hazard rates using another, larger data set, the first has zero failure rate and the second has failure rate 0.8. It can be seen that the high number of added censored observations leads to a much narrower hazard rate, with lower optimal restart time. Note that this figure is based on a different data set than the ones above, which has a larger sample size than the data set used before.

In summary, we have provided an algorithm that gives us an optimal restart time to maximise the probability of meeting a deadline only if restart will indeed help maximising that metric. So if the algorithm returns an optimal restart time we can be sure that restart will help. We found a heuristic based on the variance of the data that helps in finding the bandwidth parameter needed for the hazard rate estimator fast. We observe that small data sets usually lead to an overestimated optimal restart time. But we saw earlier (in [163]) that an overestimated restart time does much less harm to the metric of interest than an underestimated one and we therefore willingly accept overestimates. The whole restart process is automated to an extent that allows us to propose it for self-management of systems.

The runtime of the algorithm depends on the considered number of observations and on the number of iterations needed to find a good bandwidth for the hazard rate estimation. We found that for our smaller data sets with up to 400 observations less than 5 iterations are needed and the algorithm is very fast. We did not evaluate CPU time and the Mathematica implementation is not runtime optimised, but a suggestion for an optimal restart time in the above setting can be provided within a few seconds. If, however, the data set grows large, has e.g. more than 800 observations, each iteration on the bandwidth takes in the order of some 1 or 2 min. The polynomial complexity becomes relevant and the method is no longer applicable in an online algorithm.

A good heuristic for choosing the optimal bandwidth is a key part in the whole process. The better the first guess, the less iterations are needed and the faster the optimal restart time is obtained. We cannot compare our heuristic to others since in the literature nothing but pure 'trial and error' is proposed. But we can say, that for small data sets and failure probability zero the optimal restart time is obtained very fast since the heuristic provides a good first estimate of the bandwidth.

In our experience the smallest data sets were usually sufficient for a reasonably good estimate of the optimal restart time. The optimal restart time will always be placed at the end of the bulk of the observations and some few hundred observations are enough to get a notion of 'bulk' and 'end of the bulk'. If we consider that some web pages consist of up to 200 objects a data set of 100 samples is neither difficult to obtain nor unrealistic. In Internet transactions some hundred samples are easily accumulated. Furthermore, small samples seem to overestimate the optimal restart time, which does the maximised metric much less harm than underestimation.

One may argue that if everybody applies restart networks become more congested and response times will drop further. And in fact restart changes the TCP timeout – for selected applications. In our measurements we found that less than 0.5% of all connection setup attempts fail. Our method tries to detect failures faster than the TCP timeout and to restart failed attempts, since for slow connections restart typically does not lead to improved response time, whereas for failed connections in many cases it does. Failed attempts, however, are so rare that restarting those does not impose significant extra load on a network, while potentially speeding them up significantly. Obviously, if the failed attempts target at a server that is out of operation restart cannot improve the completion time. Restart can only help in the presence of transient failures. When sending requests to a permanently failed system the timeout values will increase, thus avoiding heavy load on the network connecting the failed server.

Part III
Software Rejuvenation

Introduction

Until the 1990s the common belief was that system reliability is dominated by the reliability of the hardware components a system is made of. Von Neuman addressed the issue of high-reliability already in the 1950s. He set up a model that required a redundancy factor of 20,000 to achieve a mean time between failures (MTBF) of 100 years [64, 170]. Certainly, the vacuum tubes von Neuman was thinking of were very unreliable compared to modern integrated circuits, but also his model lacked modularity. The failure of one single component would lead to a failure of the whole system, while modern systems are built of reliable modules and need much less redundancy. Still, much effort was invested in the design and construction of reliable hardware [138]. The increase in system complexity for many years exceeded the increase in component reliability. During the last decades the efforts paid off and hardware became very reliable, so that since the middle of the 1980s the functioning of software was first questioned and software failures were noticed to be a potential significant cause for system break-downs [64]. Some say that for the past two decades faster growth in system complexity kept, e.g., the processor chip failure rate almost constant. Today, hardware as well as software fault-tolerance mechanisms are necessary to achieve acceptable time between failures in the modern highly complex systems, such as, e.g., supercomputers with thousands of processors [62].

We will at first very briefly revisit general aspects of preventive maintenance. Preventive maintenance denotes a large class of methods used to prevent system failure rather than to repair and restart the failed system. Software rejuvenation is one among many methods of preventive maintenance. Preventive maintenance has already been applied in mechanical systems in the early twentieth century, while software rejuvenation is relatively new [73]. Its classification as a method of preventive maintenance with age or block replacement without emergency repair has been pointed out in [49]. Historically, preventive maintenance is concerned with hardware systems, or even manufacturing systems rather than with computing systems. Today there is no such strict distinction and in some recent publications stochastic models are called preventive maintenance models rather than software rejuvenation models to emphasise the model's universal applicability.

While preventive maintenance is a generalised concept stemming from mechanical engineering, software rejuvenation is (obviously) explicitly concerned with software in computing systems. Much of the theory of software rejuvenation discussed later in this chapter extends to technical systems in general but the system model is taylored for a computing system. The system model for software rejuvenation assumes a transaction-based software system with an underlying communication system that may use either a reliable or unreliable communication protocol.

Preventive maintenance and software rejuvenation in particular aim at preventing failures caused by faults. Not all methods of preventive maintenance are suitable for all types of failures and faults. Therefore, a classification of faults and failures is necessary. Faults are divided into permanent, or hard, faults and transient, or soft, faults. Permanent hardware faults can be circuits implementing wrong functionality due to some hardware defect which then leads to faults that can be reproduced and occur in the same way every time a circuit is executed. On the other hand, temporary faults are, e.g., bit flips due to radiation, or oscillating power. They sometimes occur, but on the next try typically cannot be reproduced. Most permanent faults are removed during testing and debugging of hardware or software, while some transient faults remain. Transient faults may not manifest in the tests and if they do so, they are not seen again and therefore are extremely difficult to detect and remove.

Hardware faults are in most cases transient faults and for hardware systems standard techniques for dealing with those exist like, e.g., checksum transmission. These techniques make sure that most transient hardware faults are either removed or worked around. It is assumed [64] that most software faults are transient as well, because most of the software faults causing permanent failures were eliminated at the latest during testing of the software. Transient software faults typically are faults related to some limit condition (counter overflow, out of memory, lost interrupt, etc.) or race conditions (semaphore problem), called Heisenbugs. Permanent faults are called Bohrbugs, after the Bohr atom, since they stay and are relatively easily detected. Software systems are assumed to be fail-fast, that means, operation of a faulty system stops and does not continue to process incorrectly. Without loss of generality we can therefore assume in our models that we know whether a system is fault-free or not.

If indeed most problems are caused by transient Heisenbugs then they can be solved by restarting the system and resuming operation of the current process. It is then highly unlikely that the same problem will occur again. The pure restart as analysed in the previous part of this thesis is applicable to treat presumably failed systems but it has not been developed explicitly as a fault-tolerance mechanism. Software rejuvenation and checkpointing both are fault-tolerance mechanisms designed to handle transient faults. For both many stochastic models have been formulated in recent years. We will not use chronological order, in which case checkpointing would be first, but instead discuss them in order of increasing complexity and start with the stochastic models for software rejuvenation. Taking checkpoints is a preventive measure but roll-back recovery follows a system failure. Checkpointing saves system states periodically during system operation, so that upon failure the system can be reset to the most recent checkpoint and no complete system restart is necessary. In contrast, software rejuvenation is purely preventive and pro-active. Software rejuvenation is based on the assumption that the process environment degrades over time and becomes faulty. The process environment is cleaned through restart, in which case, unlike with checkpointing, no process state is saved and no transactions are replayed. Software rejuvenation clearly is the simpler mechanism of the two.

Stochastic modelling of software rejuvenation is useful to determine optimal timing with respect to some metric. In preventive maintenance, depending on the replacement strategy, the timing as well as the repair action must be optimised. Software rejuvenation always uses restart of the process environment as (preventive) repair action and therefore only the timing must be determined through a stochastic model. Still, different metrics can be considered: software rejuvenation can be used to optimise system availability, the ability of a system to be operational and able to work correctly at time t, to minimise maintenance costs, or to maximise system reliability, the probability of correct operation of a system until time t. Some, but not all, of those metrics lead to equivalent optimisation problems. Cost models often are equivalent to availability models, as we will see. Both, availability and reliability are system oriented metrics and can as such be considered insensitive to the load on the system. Some of the models we discuss incorporate the load also to determine system availability. Very few models for software rejuvenation optimise job completion time, which was the key metric throughout the past chapters.

Chapter 6
Practical Aspects of Preventive Maintenance and Software Rejuvenation

In the design and development process of complex systems stochastic modelling and simulation are part of an iterative procedure. Often, before implementation the design of a new product is evaluated by means of formal and stochastic modelling. Later in the product development process prototypes are modelled, evaluated and improved. Therefore, stochastic modelling must follow or accompany the system development process. As stochastic modelling operates on a different level of abstraction than system development, the latter must not be observed in all technical detail. Technical system development issues for preventive maintenance and software rejuvenation in particular are at least as diverse as are the presented stochastic models. An exhaustive study of the history of the system development process is therefore far beyond the scope of this work. We will only briefly introduce main systems aspects of preventive maintenance as to ease later discussion of the related stochastic models.

6.1 Preventive Maintenance

Historically much earlier, but in structure very similar to software rejuvenation is the theory of preventive maintenance. Preventive maintenance actions aim at extending a systems lifetime, as opposed to repair, that would restore system operation after an outage. The bulk of work on preventive maintenance was published in the 1960s through 1980s and we will not revisit those models in detail as they typically address fault-tolerance of production systems or manufacturing systems, while recent papers are concerned with computing systems, as is this book.

To give the interested reader a starting point from where to study models for preventive maintenance in general we point to three survey articles that collect a huge amount of work in this field published before 1990. The first one [106] covers publications before 1965, the second one [120] until 1975 and the third one [160] categorises the references between 1975 and 1989, its year of appearance.

Preventive maintenance models can be categorised in different ways and to make matters more concise we will restrict ourselves to single-unit systems, since all stochastic models for software rejuvenation have been developed for monolithic systems. Multi-unit systems like the n cold drink machines mentioned in [57] require a slightly reformulated criterion of optimality.

The classification of stochastic models for preventive maintenance used in [160] and [57] suits the point of view taken in this work. The following types of models are identified.

(1) *Block replacement models* postulate complete periodic replacement of the whole unit in intervals of constant length T. In addition, failed components are removed in emergency repair (ER). Figure 6.1 shows a possible time line with periodic block replacement as well as intermediate emergency repairs.

(2) *Age replacement models* assume a lifetime distribution $F(t)$ and renewal takes place when the unit reaches age $F(t_T) = T$ or when it fails, whichever occurs first. Age replacement is more profitable than block replacement [57], when using both as cost models. It is pointed out in [57] that in practice the cost of a preplanned block replacement is less than that of an age replacement renewal.

(3) *Inspection models* assume that the state of a system that deteriorates or ages is unknown and can be learnt through inspection of the system. The purpose of these models is twofold: the required maintenance or repair action needs to be determined as well as the length of the next inspection period. Much focus is on ways to find out to what degree a system has aged or in which state of deterioration it currently is. The models use different assumptions on the nature of the aging and associate cost with inspection. They assume different degree of knowledge about the system.

(4) *Minimal repair models* assume that a single-unit system still consists of many parts. If one of the parts fails it is usually replaced (like a flat tire on a car) while leaving the aging process of the whole system unchanged. As the system deteriorates more it becomes less useful to repeatedly do minimal repair. One type of minimal repair model assumes complete periodic renewal after time intervals of length $T, 2T, 3T$, etc. which resets the failure rate. The other type of minimal repair model uses partial repair which does not reset the failure rate. Instead, in each interval $I_k = [T(k-1), Tk]$ the system failure rate $h_k(t) = h_{k-1}(t) \cdot e^{\alpha}$, where $\alpha > 0$ is a known degradation factor. Then

$$h_1(t) = h(t), \qquad h(0) \le h(t) \le h(T)$$
$$h_2(t) = e^{\alpha} \cdot h(t)$$
$$h_n(t) = e^{(n-1)\alpha} \cdot h(t)$$

○ emergency repair (ER)
● preventive maintenance (PM)

| | | | | |
| 0 | T | 2T | 3T | 4T |

Fig. 6.1 Cost based model of preventive maintenance and emergency repair

and after k partial renewals one complete periodic renewal is performed.

The many existing publications differ in assumptions on the aging, the system descriptions, the failure models, the way to determine the number k of partial renewals after which to perform a complete renewal, etc.

(5) *Shock models* assume that the failure of a system or its components is caused by shocks the system experiences. These shocks happen randomly such that the time between shocks and the damage caused by a shock are random variables that follow some probability distribution.

In practice the most commonly used preventive maintenance is block replacement, since this is the simplest strategy.

All the models listed here can be implemented as cost models, where emergency replacement after a failure (c_{ER}) has much higher cost than preventive replacement of a unit (c_{PM}), which again is more expensive than partial replacement, i.e.

$$c_{PM} \ll c_{ER}$$

Then the metric to be minimised is the total accumulated cost using the different replacement strategies while tuning the respective parameters.

Figure 6.1 displays a block replacement strategy that may have associated cost, which is not explicitly visible in the figure. There are preplanned renewals at regular intervals of length T (i.e. at time $T, 2T, 3T$,etc.) and between those unscheduled emergency repair takes place upon failure of a component. The time needed for emergency repair and preventive maintenance is irrelevant in a cost model, important is how many repairs have taken place within one preventive maintenance cycle.

Let us now derive formulas for the mean cost and the mean availability, both using block replacement. Let us first derive an approximation for the mean expected cost of block replacement. The emergency repairs (ERs) between two adjacent preventive maintenances (PMs) form a renewal process on $[0, T]$ (see Fig. 6.1 for an illustration). Let $m(t)$ be the mean number of emergency repairs (ERs) in $[0, T]$ at cost $c_{ER} = 1$ and the cost of preventive maintenance is $c_{PM} = c$, with $c < 1$, then the mean cost per unit time is [57]

$$\eta_C(T) = \frac{c + m(T)}{T}. \tag{6.1}$$

Since the renewal process often is unknown [57] proposes bounds that typically are satisfactory

$$\frac{c + F(T) + F^2(T)}{T} < \eta_C(T) < \frac{c + F(T) + F^2(T) + (F(T))^3/(1 - F(T))}{T}$$

where F is the failure time distribution and $F^n(T)$ is the $n-$fold convolution of F.

Preventive maintenance models can also be expressed as availability models. Then the steady-state availability is maximised, or the expected downtime is minimised. Instead of attaching a cost with each replacement action, the time needed

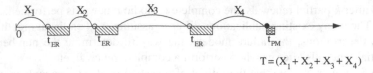

$$T = (X_1 + X_2 + X_3 + X_4)$$

Fig. 6.2 Time based model of preventive maintenance and emergency repair

for replacement is part of the model. To evaluate availability emergency repair and preventive maintenance replacement have associated duration t_{ER} and t_{PM}, respectively (see also Fig. 6.2). In analogy with the relation of the costs, we assume that

$$t_{PM} \ll t_{ER}.$$

Figure 6.2 shows a timeline with preplanned preventive maintenance and unscheduled emergency repair where for both actions a duration is defined. The time needed for repair and maintenance reduces the system operational time and hence is a punishment similar to the cost in a cost model. The operational time of the system is the remaining time between consecutive repair or maintenance. An availability model exploits the proportion of system up- and downtime due to repair and maintenance and aims at optimising the maintenance intervals in order to minimise the overall system downtime.

Supplementing (6.1) we want to maximise the steady-state availability for block replacement. The mean duration of a renewal period is $T + m(T) \cdot t_{ER} + t_{PM}$. The mean total uptime is by definition T, the mean number of emergency repairs again is $m(T)$. The total uptime is the reward in this model. This reward per total time of operation is the expected availability

$$\eta_A(T) = \frac{T}{T + m(T) \cdot t_{ER} + t_{PM}}$$
$$= \frac{1}{1 + t_{ER} \cdot \eta_C^*(T)}$$

where $\eta_C^*(T)$ is the cost for block replacement with $c = t_{PM}/t_{ER}$. So, maximising availability is equivalent to minimising the corresponding costs. We leave without proof that this holds for the other replacement strategies as well.

It is worth pointing out the relation between preventive maintenance replacement strategies and the restart discipline as defined in Chap. 4. The restart of a job has been defined using the completion time distribution. This corresponds to an age replacement strategy in preventive maintenance. Preventive maintenance is the dual action to restart in that a preventive maintenance renewal aims at extending the lifetime of a system while restarting a job aims at minimising the overall completion time of the job.

6.2 Software Rejuvenation

It is widely known that software code is never completely fault free. Industry practice indicates that carefully tested and debugged software still has at least one fault in a thousand lines of code. A recent study [68] confirmed the existence of faults in large software projects and even found the Pareto principle in general to be true, stating that 80% of all faults are located in 20% of the files, albeit not exactly these numbers. The faults and other rare coincidences, like race conditions, lead to a slow filling up of memory, reduction in available swap space, and the increasing length of the file table among other characteristics [50]. Eventually, this results in a degradation of software performance or even in a software crash-failure. The phenomenon of slowly increasing disruption of the software environment is called *software aging* [8, 50]. It has been shown to exist in web servers [103] that in fact benefit from rejuvenation.

Most software fault-tolerance and reliability techniques, such as recovery blocks [105, 126], n-version programming [7, 101, 100], N self-checking programming [105, 38] forward and backward recovery [124, 180] are reactive in nature, they include mechanisms to react to failures in order to avoid a complete system crash. Software rejuvenation, however, is a proactive technique, that aims at solving problems *before* they lead to a failure. Software rejuvenation flushes buffer queues, performs garbage collection, reinitialises the internal kernel tables, cleans up file systems, in essence: it reboots the operating environment. Since the source of the problem is unknown and cannot be removed, the only available solution is to periodically stop processing and restart the system environment and the software itself. Problems like memory overflow are then solved for the near future.

After a failure the system requires a reboot. This is an undesirable event since data might be damaged or lost if the failure happens while processing a transaction. Failures are typically rare, so a reboot due to a failure is not performed on a frequent or even regular basis. If rejuvenation is applied to prevent failures it must be done well before the software fails, and hence it is applied much more often than restart after a failure. Every unnecessary reboot of a software system causes additional downtime, which is the cost incurred by rejuvenation.

Software rejuvenation is superior to waiting for a crash and doing the reboot then only under the assumption that either

- rejuvenation is faster than the repair and restart after a failure or
- the cost of a failure is very high.

Software rejuvenation aims at increasing software reliability determined solely by the mean time to failure (MTTF) and the software availability, defined as A = MTTF/(MTTF + MTTR), where MTTR is the *mean time to repair*. Software rejuvenation does improve software availability by increasing the mean time to failure through process restarts before the software fails.

The system model assumed in the study of software rejuvenation is shown in Fig. 6.3 as a state-transition diagram [73]. Initially a system is operational (state S_0) and the probability of a failure is assumed to be negligible, therefore there is

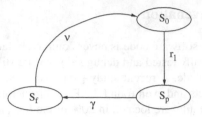

Fig. 6.3 System model without rejuvenation

no arc between the operational state S_0 and the failed state S_f. After some time the system enters the failure-probable state (state S_p), because of software aging. Then eventually the system fails (enters state S_f), is rebooted and returns to the clean state.

If rejuvenation is included in the system model, there is one additional state (state S_r), as shown in Fig. 6.4, where the system is undergoing rejuvenation.

The arcs connecting the states are labelled with (at the moment unknown) transition rates. Note, that for instance the inverse rate $\frac{1}{r_1}$ represents the average time a system is in perfect condition, called the *longevity interval* [73]. Failure and repair rate are γ and v. If the rejuvenation time is exponentially distributed its mean duration is $1/r_3$. The rejuvenation rate, when being in the failure probable state, is captured by r_2. Once the transition rates are known the expected time to failure with and without rejuvenation can be computed and the decision whether or not to perform rejuvenation can be made. Stochastic models that lead to such decisions will be investigated in the next chapter. This chapter concludes with a brief illustration of an implementation of a rejuvenation module.

Software rejuvenation has been implemented using the UNIX cron deamon [123]. Rejuvenation is implemented as an extension of the UNIX fault-tolerance tool *Watchd* [173] (in [173] furthermore, a checkpointing library is presented). *Watchd* periodically tests whether a process is still alive and restarts hanging or crashed processes. For some applications storing checkpoints is beneficial, which can be done using the library *libft*. The implementation in [73] uses the command *Addrejuv* which takes four arguments: first the process name, then a command or signal number, another signal number and the time at which to rejuvenate. *watchd* will create a script using arguments of *addrejuv* as parameters. The script is executed by the cron deamon. The commands to be executed are the second and third parameter of

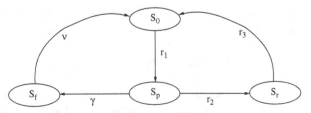

Fig. 6.4 System model with rejuvenation

addrejuv, which can be augmented by the time to wait before executing them. If no time is specified they will be executed each after waiting for 15s, finally after another 15s a SIGKILL signal will terminate the process. *Watchd* terminates and restarts an application in the same way as if there was a failure. It depends on the type of application whether the program to be restarted is kept in permanent storage or in volatile memory. In the latter case checkpointing might be necessary as well to avoid loss of data [173].

Rejuvenation has been implemented in AT&T's long distance billing system and in several regional telephone billing systems in the US. The longevity interval $\frac{1}{r_1}$ was determined before in field studies to be approximately 2 weeks. In the module only 1 week has been used as a conservative parameter. It has not been reported whether availability of the billing systems could be increased significantly.

Software availability can be improved not only by increasing the MTTF but also by applying recovery-oriented computing [17] and therewith reducing the MTTR. One can argue that in practical applications the mean time to repair is much easier to observe than the mean time to failure [44]. The former typically is measured in seconds, minutes or hours while the latter can take years. Therefore, reducing the mean time to repair is the more effective and more promising way to increase availability than prolonging the MTTF. This is the key idea behind *microreboot* [20], where only components of a software system are rebooted, not the whole system at once, leading to shorter recovery times. Microreboot is a much smaller impediment to the ongoing system operation and sometimes can be implemented such that it is not even noticed by the user. But it must be kept in mind, that microreboot aims at effectively recovering from failures after they occur and not at avoiding failures by restarting before a crash. This is an important difference to rejuvenation, where the MTTF is exploited as to restart *before* a potential failure, while microreboot aims at reducing the MTTR by partial restarts *after* a crash has happened.

The prerequisites for being able to perform a microreboot, however, are quite limiting. Microreboot can only be applied if the fault is diagnosed and associated with a component, which can be isolated from the rest of the software and can be restarted independently. Furthermore, components must be stateless and if they are not stateless to begin with all important application state must be kept in a state store. Microreboot through restart provides only application recovery while everything related to data recovery must be dealt with otherwise. The state of the observed process can be managed, e.g. by using the session state manager [95].

In most systems components will not be completely independent of each other and their structure can be represented as a tree. Figure 6.5 shows two potential tree structures. The rectangles labelled r and r_i are the restart points. The graph on the left shows a system consisting of n components, or modules $C_1, \ldots, C_n$. Components cannot be restarted individually, but only the whole system can be restarted by executing the restart point r. This is the process that is usually called software rejuvenation.

The graph on the right in Fig. 6.5 shows a software system that again consists in the components $C_1, \ldots, C_n$. Each component can be restarted individually (e.g. restarting C_1 by executing r_{111}), but sometimes the restart of a single component

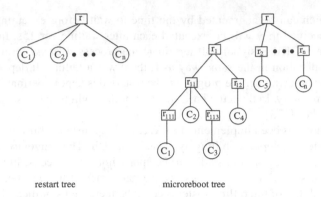

restart tree   microreboot tree

Fig. 6.5 Tree structure of global restart and restarts in microreboot systems

does not solve the problem. Then a module composed of several components is restarted, moving up one level in the microreboot tree (e.g. restarting C_1, C_2, C_3 by executing r_{11}). If this still is not sufficient the next higher level in the microreboot tree needs to be restarted until finally the system as a whole is rebooted (by executing r). Note that component C_2 cannot be restarted individually, but only together with component C_1 and C_3. In [20, 18] an implementation is presented where the components are enterprise JavaBeans and a reboot of the whole system corresponds to restarting the Java virtual machine.

Components may have different MTTR. Assume the MTTR of nodes C_1 and C_3 is much shorter than the MTTR of node C_2. If C_1 is detected to have failed one can restart only C_1 (execute restart node r_{111}). If, however, C_2 fails, both nodes C_1 and C_3 can be restarted as well without any additional time penalty. To restart C_2 the restart point r_{11} is used, which will automatically restart C_1 and C_3 too. The decision at which layer to restart components such as C_1 and C_3 and how often to do so before moving up in the restart tree is a task solved by an oracle [19]. Defining a (possibly optimal) restart strategy in a microreboot tree is a modelling issue that has not been tackled yet.

Software typically needs to be especially designed as to be amenable to restart of individual components. Microreboot has been successfully applied to the E-commerce system eBid, an auction system mimicking ebay's functionality which is implemented only using enterprise JavaBeans [20] and to the software system of Mercury, Stanford's satellite ground station [19]. Both case studies showed a remarkable decrease in mean time to repair, translating either to higher system availability or more tolerance in the fault detection time.

It should be noted that as yet no stochastic models for microreboot systems exist. One may consider the opportunistic micro rejuvenation for embedded systems [142] as closely related. This concept is a combination of software rejuvenation and micro-reboot as suitable for embedded systems. It has been analysed using a stochastic activity network with the Möbius tool [134, 31].

Furthermore, in the context of real-time systems an experimental study of component restart exists [61, 83]. In these experiments cars drive autonomously on an ellipse being directed by a controller. The controller of the first car is repeatedly being switched off and restarted as to simulate a failure and restart. The focus of the experiments lies in practical issues, such as showing that timing constraints in a real-time control system can be satisfied even when using component restart. The timing is investigated to determine experimentally whether checkpoints should be taken during system operation. The experiments do not allow for generalisation and are therefore interesting, but not of more general use.

Chapter 7
Stochastic Models for Preventive Maintenance and Software Rejuvenation

A number of stochastic models for preventive maintenance and specifically for software rejuvenation are presented and discussed in this chapter. Preventive maintenance is a method to enhance system reliability and availability. Even before a failure happens measures are taken to prevent system failure. Software rejuvenation is one such preventive action. Software rejuvenation restarts the process environment to counteract software aging. While the taken action is essentially the same as in the restart model the considered metrics are fundamentally different. In consequence, the developed models also differ from the restart model. While restart is exclusively carried out for the purpose of minimising task completion time and the system state is not considered, software rejuvenation models are in both aspects the opposite. Software rejuvenation models do not explicitly model the task completion time but instead focus on the operating environment processing the task. Software rejuvenation models minimise system downtime as well as the downtime costs not considering individual tasks. The restart model minimises task completion time while not explicitly considering possible system breakage.

In general, stochastic models can serve different purposes. They can be used to compare the performance or reliability of different system configurations, to evaluate the impact of an improvement within one system, or they can be used to formulate an optimisation problem which then helps to find an optimal parameter set. When using software rejuvenation an important question is when to rejuvenate, such that there is an overall positive effect on system performance and dependability. The optimal rejuvenation interval very much depends on characteristics of the system and the failure distribution. The models studied in this chapter aim at tuning preventive maintenance and software rejuvenation in an optimal way, but differ in their level of detail for the system description and in assumptions on probability distributions.

7.1 A Markovian Software Rejuvenation Model

We start our investigation of stochastic models for rejuvenation with the the simple model from [73] as shown in Figs. 6.3 and 6.4 on p. 128. This model has been the

K. Wolter, *Stochastic Models for Fault Tolerance*,
DOI 10.1007/978-3-642-11257-7_7, © Springer-Verlag Berlin Heidelberg 2010

first to be published and despite its simplicity and its limitations it is the basis for various other studies. In this model and its analysis the usefulness of rejuvenation is studied in terms of the downtime, or the cost associated with downtime comparing the situation without rejuvenation and with rejuvenation. An important assumption is that all state transitions are exponentially distributed. This greatly simplifies the analysis but is rarely realistic. Later work [121, 82] removes this restriction.

Let the state vector $\pi = (\pi_0, \pi_p, \pi_f)$ hold the probabilities of an application being in state S_0, S_p or S_f, respectively and let all state transition times be exponentially distributed with rates as labelled in Fig. 6.3. Failure and repair rate are γ and v, respectively and the expiry rate of the longevity interval is denoted r_1. The steady-state solution of this system is obtained as the solution of

$$\pi_0 + \pi_p + \pi_f = 1$$
$$\pi_p \cdot \gamma = \pi_0 \cdot r_1 \qquad (7.1)$$
$$\pi_f \cdot v = \pi_p \cdot \gamma$$

which evaluates to

$$\pi_f = \frac{1}{1 + \frac{v}{\gamma} + \frac{v}{r_1}}. \qquad (7.2)$$

π_f equals the proportion of downtime in the system operation, or an application runtime. If the system operation period has length L (also called *mission time*) and an associated cost c_f then the expected overall cost of an application A due to downtime is for a system without rejuvenation

$$Cost_A(L) = \frac{1}{1 + \frac{v}{\gamma} + \frac{v}{r_1}} \cdot L \cdot c_f. \qquad (7.3)$$

Including rejuvenation the system model is shown in Fig. 6.4 on p. 128, with the rejuvenation rate r_2 and the repair rate after rejuvenation r_3 being exponentially distributed as well.

The solution of the underlying Markov process with rejuvenation equals

$$\pi_p = \frac{1}{1 + \frac{\gamma}{v} + \frac{r_2}{r_3}\frac{\gamma + r_2}{r_1}} \qquad (7.4)$$

$$\pi_0 = \frac{\gamma + r_2}{r_1} \cdot \pi_p \qquad (7.5)$$

$$\pi_r = \frac{r_2}{r_3} \cdot \pi_p \qquad (7.6)$$

$$\pi_f = \frac{\gamma}{v} \cdot \pi_p. \qquad (7.7)$$

The downtime now is composed of the proportion of time the system is in the failed state plus the proportion of time it spends in the rejuvenating state $(\pi_f + \pi_r)$. Both have an associated cost $(c_f + c_r)$, so that the expected cost due to downtime of a application A^r in the model with rejuvenation evaluates to

$$Cost_{A^r}(L) = (\pi_f \cdot c_f + \pi_r \cdot c_r) \cdot L$$

$$= \frac{L}{1 + \frac{\gamma}{\nu} + \frac{r_2}{r_3}\frac{\gamma+r_2}{r_1}} \cdot \left(\frac{\gamma}{\nu} \cdot c_f + \frac{r_2}{r_3} \cdot c_r \right). \tag{7.8}$$

Note that if $r_2 = 0$ (no rejuvenation) (7.8) equals (7.3).

The parameter that needs to be tuned is the rejuvenation rate r_2. Whether the downtime or the cost of downtime will increase or decrease when changing r_2 can be seen in the sign of the derivative of the downtime with respect to r_2. The derivative of $downTime_{A^r}(L)$ with respect to r_2 evaluates to

$$\frac{d}{dr_2} downTime_{A^r}(L) = L \cdot \frac{\gamma}{\nu r_1 r_2} \cdot \frac{1}{\left(1 + \frac{\gamma}{\nu} + \frac{\gamma}{r_1} + \frac{r_2}{r_1} + \frac{r_2}{r_3}\right)^2} \cdot \left(\nu(1 + \frac{r_1}{\gamma} - r_3)\right).$$

$$\tag{7.9}$$

It should be pointed out that the above derivative changes its sign independently of r_2 which means that either immediate rejuvenation $(r_2 = 0)$ or no rejuvenation $(r_2 = \infty)$ will be optimal.

The denominator of (7.9) is always positive and hence the sign of (7.9) only depends on the last term. The last term relates the time to rejuvenate $(\frac{1}{r_3})$ with the failure and repair rate. More precisely, if

$$r_3 < \nu + \frac{r_1\mu}{\gamma} \qquad \text{then} \qquad \frac{d}{dr_2} downTime_{A^r}(L) > 0 \tag{7.10}$$

then increasing the rejuvenation rate increases the downtime, while if

$$r_3 > \nu + \frac{r_1\nu}{\gamma} \qquad \text{then} \qquad \frac{d}{dr_2} downTime_{A^r}(L) < 0 \tag{7.11}$$

increasing the rejuvenation rate decreases the expected downtime. In the latter case rejuvenation should be immediately performed as the system enters the failure probable state.

It is intuitively evident that an application benefits from rejuvenation only if rejuvenation is faster and less expensive than repair. This can be seen in the above condition as well as in the condition on the downtime cost below. Clearly, if the recovery after rejuvenation and the repair after a failure take equally long (i.e. $\nu = r_1$), then

$$\nu(1 + \frac{r_1}{\gamma}) - r_3 = \frac{\nu r_1}{\gamma}, \tag{7.12}$$

which is always positive, independent of all other parameters and rejuvenation will always increase the system downtime (as long as $r_2 > 0$).

The derivative with respect to the rejuvenation rate r_2 of the total downtime cost (Eq. (7.8)) equals

$$\frac{d}{dr_2} Cost_{A^r}(L) = L \cdot \frac{1}{v r_1 r_3} \cdot \frac{1}{(v r_1 + \gamma v + \gamma r_1)} \cdot \frac{1}{\left(1 + \frac{\gamma}{v} + \frac{\gamma}{r_1} + \frac{r_2}{r_1} + \frac{r_2}{r_3}\right)^2} \cdot \left(c_r - c_f \frac{\gamma(r_1 + r_3)}{\gamma(v + r_1) + v r_1}\right). \quad (7.13)$$

Again, the sign of the above expression purely depends on the last term, which is independent of the rejuvenation rate r_2.

Figure 7.1 on the next page shows plots of the two terms which determine the sign of the derivative of the downtime and the derivative of the downtime cost both with respect to the rejuvenation rate. The zero plane is added to the graph to separate the positive from the negative parts of both curves. Some parameters had to be set. The longevity is set to 10 days ($1/r_1 = 10 \times 24$), the MTBF (mean time between failures)[1] is 1 year ($1/\gamma = 12 \times 30 \times 24$), the cost of rejuvenation c_r equals 5, and the cost of repair after a failure c_f equals 50. Those numbers are chosen similar to the parameters in [73]. It is then possible to observe the impact of the MTTR ($1/v$) and the time to rejuvenate ($1/r_3$) on the derivative of both the downtime and the downtime cost. The figures represent only the last term in (7.9) and (7.13), being responsible for the sign of the equation. Important in the figures is therefore whether the curves are greater, equal or less than zero, but the absolute values have no meaningful interpretation.

If both v and r_3 are equal to zero, the time to repair is infinite and increasing the rejuvenation rate does neither reduce nor increase the total downtime. The derivative equals zero, since its last term equals zero. Note that the graph for the change in downtime cost uses the same parameters but is plotted in a different range for better visibility. For MTTR $= \infty$ the derivative of the downtime cost converges to minus infinity, meaning that rejuvenation reduces the cost of downtime dramatically, for without rejuvenation the system will be down for ever.

Two conclusions should be drawn from the two graphs in Fig. 7.1: (1) values exist for the MTTR and for the repair rate after rejuvenation such that both the downtime and the downtime cost may decrease or increase with changing rejuvenation rate. And (2) for some choices of the two parameters both downtime and downtime cost are reduced, but also parameter values exist such that one is reduced and the other increased. I.e. at $r_3 = 2$ and $v = 0.3$ cost decreases with changing rejuvenation rate, but the downtime does not.

---

[1] Note that usually the MTBF is the sum of the MTTF (mean time to failure) and the MTTR (mean time to repair), making a specification of the repair rate obsolete. In [73], however, both are given, so perhaps MTTF is meant instead of MTBF.

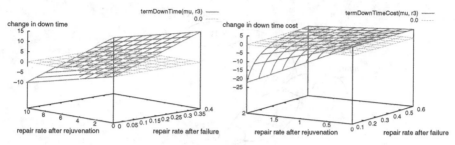

Fig. 7.1 Derivative of the downtime and the downtime cost with respect to the rejuvenation rate

It is worth mentioning again, that in this model the downtime or downtime cost can be optimised, if at all, only by the extreme parameter choices either to rejuvenate immediately or never. It has been pointed out in [34, 35] and in Sect. 3.2 that an exponentially distributed restart time, or rejuvenation interval, will always imply that the optimal restart or rejuvenation interval has length zero. Rejuvenation should be performed as soon as the model enters the failure probable state, an observation that is supported by the results obtained in [73]. Therefore, later work in [34] extends the cost model in this section and [35] extends the availability model such that all state transitions can be other than exponentially distributed. Both papers also investigate a slightly different state transition model and provide a non-parametric, data-driven method for experimentally obtaining the optimal rejuvenation interval.

In summary, the very simple model in [73] gives much insight in the interplay of rejuvenation and repair after failure of a system. The model, however, only incorporates a very simple form of aging, the notion of a longevity period, which has an exponentially distributed duration and is, therefore, not amenable for detailed analysis.

7.2 Aging in the Modelling of Software Rejuvenation

The system model in the previous section has an underlying homogeneous Markov process. The longevity period represents the concept of aging in a rather unnatural way. The model in this section represents aging using a time dependent failure rate that increases over time, representing an increasing likelihood of system breakdown. This leads to the simpler system model shown in Fig. 7.2 and is the more common way today to model component exhaustion.

While the model in the previous section solely considered a software system, the one in this section also includes arrival and service of jobs as well as queueing. The analysis of the model consists of two parts: first the system model is analysed and then the queueing model. First a system metric, the system availability, is computed and then the loss probability, which is a job related metric.

The necessary parameters of the model are on one hand the arrival and service characteristics and on the other hand failure, repair and rejuvenation parameters.

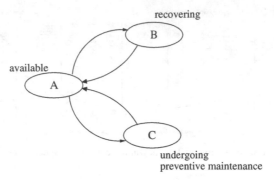

Fig. 7.2 System model for preventive maintenance and failure/repair

The time when to trigger rejuvenation is being optimised through evaluation of two different policies.

- Policy **I** is to rejuvenate after fixed time intervals of length δ, while in
- Policy **II** after waiting δ time rejuvenation is only performed when the transaction queue is empty.

An optimal choice of δ such that system availability is maximised and the loss probability is minimised formulates the optimisation problem of this section. Presumably the model is too complex to perform direct optimisation. In [48, 153, 49] the strategies are evaluated in the same model with different parameter sets showing the best parameter set. How to parameterise the model in a realistic way using the data collection from [159] is shown in [153].

Let us now formally define the model parameters. We assume a failure rate $\gamma(t)$, where the time to failure X is distributed as $F_X(t) = 1 - e^{-\int_0^t \gamma(s)ds}$. In [48] the failure rate can be time dependent, while in [153] it may also depend on the system load $L(t)$, or on the actual processing time [49, 153]. The processing time is defined as

$$L(t) = \int_{\tau=0}^{t} \sum_i c_i p_i(\tau) \, d\tau$$

where p_i denotes the probability of i jobs waiting in the queue and c_i expresses the degradation level of the software with i jobs waiting.

The failure rate γ can depend on a combination of both time and load as well as on the mean accumulated work [49]. This allows to capture aging behaviour as well as overload problems. Note that the time to failure is not in general exponentially distributed, making the model in Fig. 7.2 a Markov regenerative process (MRGP) [87], or Markov renewal process with state space $Z(t) = \{A, B, C\}$ [29] and no longer a continuous-time Markov chain (CTMC) [78]. The service rate $\mu(.)$ is defined similarly. It can be a function of time $\mu(t)$, or a function of the system load $\mu(N(t))$, where $N(t)$ is the number of transactions waiting in the queue to be served, or of both. The service rate is always measured from the last renewal time of the software and the service time has distribution $1 - e^{-\int_r^t \mu(\cdot)ds}$ if service starts at time r. Transactions arrive at constant rate λ to the software system which has a queueing capacity of K transactions or jobs.

Let us assume general distributions F_{D_f} and F_{D_r} for the random variables describing the time to recover from a failure (D_f) and the time to recover from rejuvenation D_r. No further assumptions on the distributions are necessary, only that their respective expectations $E\left[D_f\right]$ and $E[D_r]$ are finite.

If $\{Z(t), t \geq 0\}$ denotes a stochastic process representing whether the software system is in one of the three states A, B, or C, and the sequence of random variables $S_i, i > 0$ represent the times at which an application moves from one state to the next, then the entrance times S_i constitute renewal times and $\{Z(t), t \geq 0\}$ is an embedded discrete time Markov chain (DTMC) [29, 87]. The embedded DTMC has probability matrix $\mathbf{P}$ and steady-state probability vector $\pi = (\pi_A, \pi_B, \pi_C)$ with

$$\mathbf{P} = \begin{pmatrix} 0 & p_{AB} & p_{AC} \\ 1 & 0 & 0 \\ 1 & 0 & 0 \end{pmatrix}.$$

The solution of the homogeneous linear equation $\pi(\mathbf{I} - \mathbf{P}) = \mathbf{0}$ gives the steady-state probabilities of the DTMC being in one of the three states. The solution evaluates to

$$\pi = \left(\frac{1}{2}, \frac{1}{2}p_{AB}, \frac{1}{2}p_{AC}\right). \tag{7.14}$$

The expected availability of the software system can now be computed to be the average proportion of time the system is in state A. Let U be a random variable denoting the sojourn time in state A, the expected sojourn times in states B and C are defined above as $E\left[D_f\right]$ and $E[D_r]$ respectively. Then the expected steady state availability A_{ss} of the software system is

$$A_{ss} = \frac{\pi_A E[U]}{\pi_B E\left[D_f\right] + \pi_C E[D_r] + \pi_A E[U]} = \frac{E[U]}{p_{AB} E\left[D_f\right] + p_{AC} E[D_r] + E[U]} \tag{7.15}$$

To be able to compute system availability the three parameters p_{AB}, p_{AC} and $E[U]$ must be known. They all depend on which rejuvenation strategy is applied and are therefore for both cases defined below.

A second metric of interest is the loss probability for jobs waiting for service as the system fails or is stopped for rejuvenation. To compute the loss probability the queueing behaviour must be defined. Only in state A a queue of waiting jobs $N(t)$ exists. In both states B and C the queue is empty and $N(t) = 0$ since all jobs arriving while the system is in the failed state or being rejuvenated are lost by definition. The full description of the considered stochastic process is $\{Z(t), N(t)\}$. This process is a MRGP since the time behaviour of $N(t)$ changes while the system is in state A. This is because the service rate $\mu(.)$ may be time dependent to model degrading service quality.

The expected number of lost jobs is composed of three quantities:

- expected number of jobs lost because of failure and initiating rejuvenation
- expected number of jobs discarded during recovery from failure and rejuvenation
- expected number lost because the buffer is full.

Formally for Policy **II**:

$$
\begin{aligned}
P_{\text{loss}} &= \frac{\pi_A \text{E}[N_l] + \lambda \left(\pi_B \text{E}[D_f] + \pi_C \text{E}[D_r] + \pi_A \int_0^\infty p_k(t)\, dt \right)}{\lambda \left(\pi_B \text{E}[D_f] + \pi_C \text{E}[D_r] + \pi_A \text{E}[U] \right)} \\
&= \frac{\text{E}[N_l] + \lambda \left(p_{AB} \text{E}[D_f] + p_{AC} \text{E}[D_r] + \int_0^\infty p_k(t)\, dt \right)}{\lambda \left(p_{AB} \text{E}[D_f] + p_{AC} \text{E}[D_r] + \text{E}[U] \right)}
\end{aligned}
\tag{7.16}
$$

where $\text{E}[N_l]$ is the expected number of jobs in the buffer when the system exits state A. The expected number of jobs that are lost due to failure of the system or initiation of preventive maintenance depends on the rejuvenation strategy and will therefore be defined below.

Using Policy **I** the upper limit of the integral in (7.16) is δ the rejuvenation interval (instead of ∞). After a period of length δ the system always leaves state A and the buffer is flushed.

W denotes the mean total time the transactions spend in the system while the software is in state A

$$
W = \int_{t=0}^\infty \sum_i i p_i(t)\, dt.
$$

Time W consists of the time spent in the system by jobs which were actually served (W_S) and the time spent in the system by those that were discarded at some point (W_D). Of interest is only the former. But the sum of both $W = W_S + W_D$ can be used as an upper bound.

The response time can then be bounded by

$$
T_{res} < \frac{W}{E - \text{E}[N_l]}
\tag{7.17}
$$

where E is the mean number of transactions accepted for service while the system is in state A. E is composed of the total number of transactions arriving to the system minus those that find the buffer already full

$$E = \lambda \left(\mathrm{E}[U] - \int_{t=0}^{\infty} p_K(t) \, dt \right).$$

Out of these $\mathrm{E}[N_l]$ are still discarded later, because the system fails or rejuvenation is initiated.

7.2.1 Behaviour in State A under Policy I

The three metrics of interest, the steady-state availability, the loss probability and the upper bound on the response time, are computed separately for both rejuvenation policies. As the queueing process $N(t)$ is different for both policies and is needed to compute the metrics of interest, it has to be studied for both policies separately. In Policy **I** the system remains in state A until it fails, or until δ time units have elapsed and rejuvenation is initiated, whichever comes first. For $Z(t) = A$ the subordinated process until the next regeneration point is the queueing process $N(t)$, shown in Fig. 7.3. The queueing process is determined by arrivals of transactions to the system and their being processed. It corresponds to a birth-death process with an added absorbing state. At time $t = \delta$ the system moves to one of the absorbing states $(0', \ldots, K')$. This is not reflected in the figure.

The states $0', \ldots, K'$ are needed for the computation of metrics, otherwise they could be lumped into one absorbing 'down' state.

The system of equations to be solved form a subset of the equations for Policy **II** given in (7.19), where the separate equations for $p_1(t)$ and $p_1'(t)$ are omitted and i always ranges from $1, \ldots, K$.

Once the solutions $p_i(t), 0 \le i \le K$ and $p_i'(t), 0' \le i' \le K'$ are available the following terms can be computed in a straightforward manner:

$$p_{AB} = \sum_{i=0'}^{K'} p_i(\delta)$$

$$p_{AC} = 1 - p_{AB}.$$

Then, the steady-state availability (7.15) can be computed.

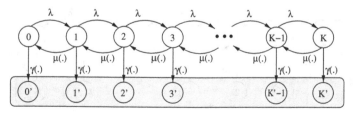

Fig. 7.3 Subordinated non-homogeneous CTMC for $t \le \delta$

The expected sojourn time in state A is

$$E[U] = \int_{t=0}^{\delta} \left(\sum_{i=0}^{K} p_i(t) \right) dt$$

and

$$E[N_l] = \sum_{i=0}^{K} i(p_i(\delta) + p_i'(\delta)),$$

which allow to compute the loss probability and the upper bound on the response time as given in (7.16) and (7.17).

7.2.2 Behaviour in State A under Policy II

Using Policy **II** makes matters more complicated, since the system behaviour changes when δ time units have passed after each renewal. While the system is in State A it is in normal operation until time δ after which no more arriving jobs are accepted until the system is empty and being rejuvenated.

Therefore, for $t \le \delta$ the system behaves as with using Policy **I**, and as depicted in Fig. 7.3, while as $t > \delta$ it behaves differently. The subordinated non-homogeneous CTMC when $t > \delta$ is shown in Fig. 7.4.

State 0 now belongs to the absorbing states for when the system is empty the subordinated process is left for rejuvenation. No arriving transaction is accepted when $t > \delta$ and the queue is empty.

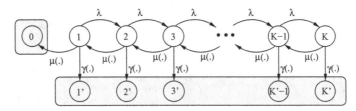

Fig. 7.4 Subordinated non-homogeneous CTMC for $t > \delta$

The system of equations to be solved is

$$\frac{dp_0(t)}{dt} = \mu(.)p_1(t) - (\lambda' + \gamma(.))p_0(t)$$

$$\frac{dp_1(t)}{dt} = \mu(.)p_2(t) + \lambda' p_0(t) - (\lambda + \mu(.) + \gamma(.))p_1(t),$$

$$\frac{dp_i(t)}{dt} = \mu(.)p_{i+1}(t) + \lambda p_{i-1}(t) - (\lambda + \mu(.) + \gamma(.))p_i(t), \quad 2 \leq i < K \quad (7.18)$$

$$\frac{dp_K(t)}{dt} = \lambda p_{k-1}(t) - (\mu(.) + \gamma(.))p_K(t)$$

$$\frac{dp_{0'}(t)}{dt} = \gamma'(.)p_0(t)$$

$$\frac{dp_{i'}(t)}{dt} = \gamma(.)p_i(t), \qquad 0 \leq i' \leq K$$

where

$$\lambda'(t) = \begin{cases} \lambda & \text{if } t \leq 0 \\ 0 & \text{else} \end{cases} \quad \text{and} \quad \gamma'(t) = \begin{cases} \gamma & \text{if } t \leq 0 \\ 0 & \text{else.} \end{cases}$$

The initial condition is

$$p_0(0) = 1, \quad p_i(0) = 0, \quad 1 \leq i \leq K, \qquad p'_i(0) = 0, \quad 0' \leq i' \leq K'. \quad (7.19)$$

If the arrival rate to the queue and the service rate of the queue depend on the queue length, i.e. $\mu(.) = \mu(L(t))$ and $\gamma(.) = \gamma(L(t))$ then one additional equation is needed to determine the solution of the process

$$\frac{dL(t)}{dt} = \sum_{i=0}^{K} c_i p_i(t). \tag{7.20}$$

The probability p_{AB} is computed by solving the above system of equations at $t = \infty$ and it equals

$$p_{AB} = \sum_{i=0'}^{K'} p_i(\infty)$$

and therefore

$$P_{AC} = 1 - P_{AB} = p_0(\infty)$$

because only for the absorbing state a solution greater than zero at $t = \infty$ exists. The mean sojourn time in State A is

$$
\mathrm{E}[U] = \int\limits_{t=0}^{\delta} \left(\sum_{i=0}^{K} p_i(t) \right) dt + \int\limits_{t=\delta}^{\infty} \left(\sum_{i=1}^{K} p_i(t) \right) dt
$$

$$
= \int\limits_{t=0}^{\delta} p_0(t) \, dt + \int\limits_{t=0}^{\infty} p_0(t) \left(\sum_{i=1}^{K} p_i(t) \right) dt.
$$

The mean number of transactions in the queue when State A is left equals

$$
\mathrm{E}[N_l] = \sum_{i=0}^{K} i p_{i'}(\infty).
$$

Now the steady-state availability (7.15), the loss probability (7.16) and the upper bound on the response time (7.17) can be calculated for Policy **II**.

Results taken from [45] are shown in Fig. 7.5. The used parameters are a failure rate that is defined as the hazard rate of a Weibull distributed random variable, using the MTTF of 240 h for defining the distribution's parameters. The MTTR equals 0.85 h, the service rate is variable with a maximum of $15/h$ and a minimum of $5/h$, the arrival rate is $6/h$ and the buffer size is 50. The expected time needed for rejuvenation is varied across the curves for both Policy **I** and Policy **II**.

In Fig. 7.5 each curve shows either the availability or the loss probability for different values of the rejuvenation interval δ. The curves are labelled with the policy and the expected rejuvenation time.

The most important interpretations of the results in Fig. 7.5 are as follows.

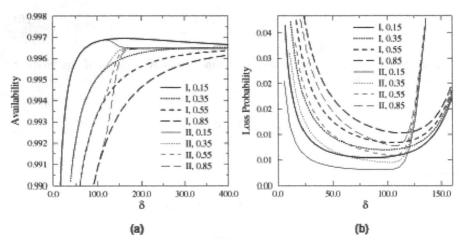

(a)   (b)

Fig. 7.5 Availability and loss probability taken from [49]

- If the primary aim is to minimise loss probability, Policy **II** is always better than Policy **I**, while for maximising availability this is not obvious from the given results.
- The value of δ that minimises the loss probability is lower than the value of δ that maximises availability [153, 49].
- If the MTTF is exponentially distributed there exists no rejuvenation interval or rejuvenation wait δ that optimises steady-state availability, while the loss probability can still be minimised [49].

An obvious observation is that the longer the rejuvenation time the higher is the loss probability and the lower is the availability.

Apart from the characteristics of the results themselves it is of interest how this model compares with the restart model discussed in Chaps. 4 and 5 and with the Markovian model presented in the previous section.

Let us compare the latter first. There are many differences and not too many similarities between this model, and its results, and the one in the previous section. First of all, Fig. 7.5 shows optimal rejuvenation interval lengths other than zero or infinity. The models are so different in structure that it is not even possible to say whether they would lead to the same conclusion for the same parameter set. But it can be said that clearly the model in this section includes much more detail and therefore provides more detailed results while being less straightforward to solve and interpret.

Certainly, including the notion of *jobs* or *transactions* adds much value, since not only system availability but also other important metrics such as loss rate and response time can be considered. The queueing model new in this model adds a customer perspective to the pure system view of the problem. The restart model in Chaps. 4 and 5 completely hides the system view in the completion time distribution (which of course in some way depends on the system dynamics) and only exhibits a customer view. It cannot be judged from the material in the given references whether the non-exponential repair does improve the model quality. Applying the different approaches to a situation with an identical parameter set would certainly answer some open questions and allow us to judge as to whether or not one model gives the more conservative results compared with the others.

Figure 7.5 indicates that when minimising the loss probability in the non-Markovian model one would chose the rejuvenation interval δ rather too small than too large but for maximising availability it is better to choose δ rather too large. Similar conclusions were drawn in Sect. 4.2.1 for improving moments of completion time through restart. The rejuvenation interval with respect to availability therefore behaves analogously to the restart interval to minimise completion time, which is rather chosen a little longer (see Fig. 4.3), while the rejuvenation interval with respect to the loss probability shows the opposite behaviour.

7.3 A Petri Net Model

Petri nets are a modelling formalism well-suited and often-used for describing concurrency and synchronisation in system actions [3]. Reformulating the rejuvenation model from the previous section as a Petri net model allows for much more modelling detail as was shown in [45]. We present a modified version of the model in [45] here. The model in [45] is a stochastic reward net (SRN), which is similar to a GSPN as in [3]. In SRNs reward definitions are an integral part of the model, arc cardinality can be marking dependent and transitions can be enabled by state-dependent functions. The added structural components can make models more compact, but they do not increase the modelling power of the formalism.

The model we present here belongs to the class of DSPNs [55, 56], since activities in the model can have other than exponentially distributed duration. The rejuvenation model only uses deterministic timing for modelling the clock which initiates rejuvenation after the rejuvenation interval of fixed length has expired.

The DSPN model in Fig. 7.6 roughly corresponds to the rejuvenation model from the previous section using Policy **I**. Rejuvenation is carried out whenever the rejuvenation interval expires, which is modelled as the firing of transition Tclock. In the experiments the firing time of Tclock is varied to find the optimal rejuvenation interval.

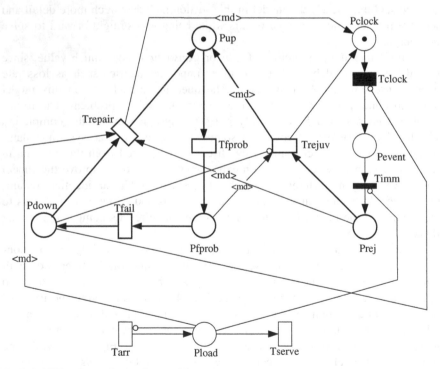

Fig. 7.6 DSPN of the rejuvenation model

The solid lines in the Petri net model indicate the part of the model that is directly translated from the state transition diagram in Fig. 7.2, the subordinated CTMCs in Figs. 7.3 and 7.4 on p. 142 representing the jobs arriving to the system and leaving the system are here directly included in the model, represented by the place Pload and the transitions Tarr and Tserve. The Petri net model implements Policy **II** as this is the better of both policies with respect to most criteria. The customer queue is preempted upon repair of the system. In [45] the customer queue is also preempted while rejuvenation takes place. The corresponding arc is omitted in our model, since it showed no effect on the results. Since rejuvenation is only initiated when the queue is empty one would not expect preemption to happen frequently.

The clock measuring the time intervals after which rejuvenation is initiated is modelled using a transition with deterministic firing time (Tclock). Its firing time is the parameter to be tuned. In [45] the deterministic firing time is approximated by a series of 10 transitions with exponentially distributed firing times, together representing an Erlang distributed activity. Luckily, the software tool TimeNET allows for generally distributed firing times, deterministic ones in particular, hence making the analysis more convenient and much faster. All other activities are exponentially distributed.

The parameters of our model are identical to those in [45] – if they were specified. In particular they are the following long term averages: Customers arrive in a Poisson process with rate 30 (here we selected only one of the cases studied in [45] and simplified the model in that we omitted the incompletely specified modulating process) and the jobs require a service time of 2 min. The queue is limited to just one customer by the inhibitor arc connecting place Pload and transition Tarr. The system stays in the robust state for 10 days and after reaching the failure probable state it fails in another 30 days. Rejuvenation takes on the average 10 min and recovery after a failure takes 200 min. In [45] rejuvenation and failure recovery take equally long, in which case rejuvenation cannot be beneficial (it would always pay off to wait for a failure), so the results provided in [45] cannot be obtained using the parameter values as given in that paper.

The marking dependent arcs assure that a token is removed from a place only if the place holds a token, making sure that the transition can fire even if the input place is empty. Similarly, the marking dependent arcs connecting transition and place make sure that a token is deposited only if the place is empty. This happens if the system is being rejuvenated while being in the failure probable state or if the system fails during rejuvenation and then undergoes repair instead of rejuvenation.

The parameters used to solve the model in Fig. 7.6 are listed in Table 7.1 (in rates per hour).

Table 7.1 Parameter values used for the DSPN

Tfprob	$4.166 \cdot 10^{-3}$	Trejuv_1	0.3	Trejuv_2	6
Tfail	$1.388 \cdot 10^{-3}$	Tarr	30		
Trepair	0.3	Tserve	30		

The model is solved using the software tool TimeNET [182] using the steady-state analysis module for increasing rejuvenation interval from 50 to 700 in steps of 10. The runtime for this experiment was 36.7148 s. Two different parameter sets were evaluated, the first one being as described in [45] with the rejuvenation time and the repair time being identical. The results are shown in Fig. 7.7 on the left. Obviously, the longer the rejuvenation interval the less jobs are lost due to rejuvenation and the more jobs are lost due to system failure. But the number of jobs lost due to system failure soon converges to a constant and the optimal rejuvenation interval is infinitely long. This model does not incorporate cost and hence the number of jobs lost is the metric to be minimised. As mentioned above, if rejuvenation and repair are associated with the same expected number of lost customers, there is no benefit in rejuvenating before a failure happens. This statement is supported by the results from the model.

On the right hand side in Fig. 7.7 are the expected loss rates from the same model, the only difference being that the mean repair time is much longer than the rejuvenation time. In this case a clear optimum for the rejuvenation interval exists, such that the total expected number of jobs lost is minimised.

Policy **I** from the previous section has been implemented by cutting the inhibitor arc from place Pload to transition Timm, hence allowing for rejuvenation even if there is a job being processed. The experiment included solutions of the model for the rejuvenation interval ranging from 50 to 500 in steps of 10. The runtime for the experiment was 26.5997 s. The result can be seen in Fig. 7.8. Surprisingly, the curves are almost identical to those in Fig. 7.7 on the right. The optimal rejuvenation interval and the corresponding expected loss rates differ only in the fifth digit after the decimal point.

The conclusion from the Petri net model must be, that an optimal rejuvenation interval exists if repair takes sufficiently longer than rejuvenation. The choice of the policy is less important. The model allows to determine whether rejuvenation is beneficial for given rejuvenation and repair time. If rejuvenation improves the steady-state system availability and loss rate the optimal rejuvenation interval can be determined.

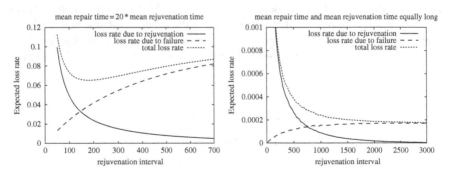

Fig. 7.7 Expected number of jobs lost versus rejuvenation interval using Policy **II**

More Petri net models of software rejuvenation exist and in [47] a similar improvement as we did here was made. The rejuvenation model is a Markov regenerative stochastic Petri net (MRSPN), which has an underlying stochastic process that is a Markov regenerative process. In consequence also actions with deterministic timing can be included in the model. This is the main difference between the Petri net models in [45] and [47]. The models, furthermore, differ in minor details implementing the reset of the system after rejuvenation. Still, the two models cannot be directly compared since [47] uses the model parameters such as MTTF and rejuvenation interval from [73] and computes different metrics such as expected downtime, transient expected unavailability and expected cost.

Figure 7.9 shows the transient expected downtime due to rejuvenation, failure and both. The left plot presents the transient results until time 6,000 to demonstrate the slow convergence of a model with mixed Markovian and deterministic behaviour. The right plot shows the same results. For clearer presentation only the first 600 time units are displayed. It takes the model several thousand time units to reach steady-state. Transient availability in the SRN model in [47] is in steady state already after 1,500 time units. This might be caused by a slightly different parameter set, but also Markovian models reach steady state much faster than models with deterministic timed activities.

In Fig. 7.9 the downtime increases over time and is reset to a much lower value by rejuvenation, taking place in regular intervals. Unavailability due to rejuvenation, on the other hand, initially happens almost with probability upon expiry of the rejuvenation interval. Only slowly and in a very long term average the probability of rejuvenating decreases and hence the downtime then is less dominated by the downtime due to rejuvenation. Notice the logarithmic scale of the ordinate and the sharp peaks at the rejuvenation times.

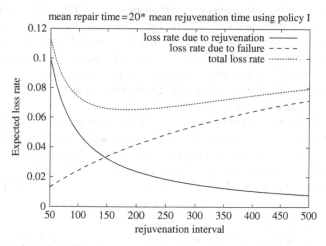

Fig. 7.8 Expected number of jobs lost versus rejuvenation interval using Policy **I**

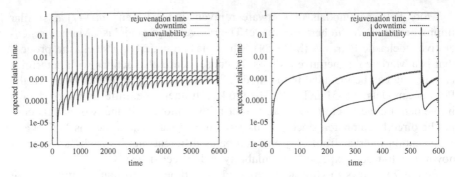

Fig. 7.9 Transient downtime of the system due to rejuvenation, failure or both

A cable modem termination system is modelled as a stochastic reward net in [96] and cluster systems in [157, 21], leading to very complex, application specific models. Time based as well as measurement based rejuvenation is being modelled. These models include a complex structure and interdependencies of potential faults but the timer counting the rejuvenation interval again is approximated using an Erlang distribution. The computed metrics are capacity oriented availability and the downtime cost.

The Petri net model for software rejuvenation can include much detail and structural dependencies, which clearly is a great advantage in comparison with other modelling formalisms. It allows for very complex structures, which then are extremely hard to debug. But the main disadvantage of a Petri net model is that parameter optimisation must be done 'by hand' in carrying out sequences of experiments.

Software rejuvenation has also been modelled using fluid stochastic Petri nets [12]. Extending stochastic Petri nets with fluid places, that hold a continuous amount of fluid, rather than discrete tokens was first proposed in [155], solution methods for these models were presented in [72, 28]. The formalism was extended to include variability in the movement of fluid [175], applications and solution quality of the formalism were studied in [179]. Further extensions introduced jump arcs [176], or flush-out arcs [66, 65], which have similar semantics. Jump arcs remove or deposit a certain amount of fluid at once, instead of continuously as the fluid arcs do while flush-out arcs remove all content of a fluid place at once.

The fluid stochastic Petri net models in [12] represent a degrading software system with failure and repair that is periodically being checkpointed[2] as well as rejuvenated. Four fluid places are used to represent the level of degradation of the system, the unsaved work performed since the last checkpoint and two fluid places represent the time since the last renewal (checkpoint, rejuvenation, or failure). The work as

---

[2] For an elaborate discussion of checkpointing and stochastic models for checkpointing see the next chapter.

well as the clocks are reset upon failure, rejuvenation, or checkpointing. No general purpose Petri net tool is used for solving the model. Instead, a discretisation scheme has been implemented for this particular model. The computed metrics are on one hand the relative time spent in the discrete states denoting the mode of operation of the system (normal operation, failed state, rejuvenating, etc.) and the proportion of useful work performed until time τ, called work efficiency.

The results show similar saw-tooth patterns as the ones in Fig. 7.9, although the peaks are not as sharp and especially the work efficiency is quickly smoothened by the effects of exponentially distributed activities in the model. The fluid stochastic Petri net model is used to show characteristics of the model, there is no parameter tuning or optimisation being carried out. Neither are the effects of checkpointing versus rejuvenation evaluated. The model serves a purely descriptive purpose.

7.4 A Non-Markovian Preventive Maintenance Model

Lately, models have been named *preventive maintenance models* to emphasise their general applicability to technical systems and not being tailored for a special kind of system. The non-Markovian preventive maintenance model is taken from [13] where it serves as an example to illustrate the modelling power of the non-Markovian model class *Markov-regenerative stochastic Petri nets (MRSPN)*. As it is unclear which software tool has been used to obtain the results presented in [13] we will use the Möbius modelling environment to specify and analyse the model. Figure 7.10 shows a stochastic activity network (SAN) [134] as an atomic model in Möbius [31].

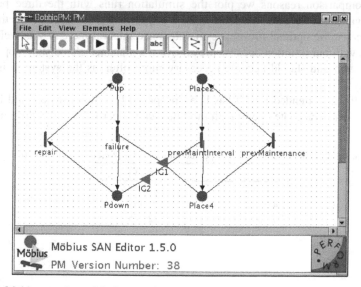

Fig. 7.10 Möbius atomic model of preventive maintenance

Unlike the models in the previous sections this model does not include a queue. It solely consists of a system model. The model represents a system that can be in one of two states, either it is operational (up) or failed (down). Added is a preventive maintenance unit with a timer. When reaching the end of the preventive maintenance interval the timer expires and the next preventive maintenance action is triggered. The activity representing preventive maintenance preempts the activity *failure* and hence the failure process. In consequence the model acts *as new* after the preventive maintenance action has been completed. The two input gates (IG1 and IG2) are implemented as inhibitor arcs to disable failure during preventive maintenance and preventive maintenance during repair. The activity *failure* models the aging of the system by being distributed according to a Weibull distribution[3] with scale parameter λ and shape parameter $\alpha = 2.0$ which produces an increasing hazard rate and hence models aging of the system (Table 7.2).

In [13] the analytical solution of the underlying semi-Markov process (SMP) is derived and used for obtaining the desired results. We will use here the simulation component of the Möbius modelling tool to solve the model in Fig. 7.10. The parameters are the following. Initially the system is up. It fails according to a Weibull distribution with shape parameter 2.0 and varying scale parameter λ as shown in Table 7.2. The repair time is exponentially distributed and takes on the average 10 h. The preventive maintenance interval δ is varied as shown in Fig. 7.11 and preventive maintenance as such takes 1 h. Note that in [13] the claim is that the Weibull distribution with parameters $\alpha = 2.0$ and $\lambda = 2 \cdot 10^{-7}$ would have an expectation of 1981.66. We could not recompute the first moment at these parameter values and therefore changed the value of λ in order to keep the same MTTF. In doing so we find similar optimal preventive maintenance intervals as presented in [13].

For comparison reasons we plot the simulation runs with the three parameter configurations as given in Table 7.2. We find that as the MTTF of the system increases the optimal preventive maintenance interval increases as well. This complies with our intuition because preventive maintenance also disturbs operation of the system and should become necessary more often as the system fails more frequently.

Recently, an extended model has been proposed [82] that uses a stand-by unit and can hence rejuvenate one or both modules, called partial or full rejuvenation.

Table 7.2 Parameter values λ and expected values of the Weibull distribution ($\alpha = 2$)

λ	$E[X]$
$0.5 \cdot 10^{-3}$	1772.45
$0.455 \cdot 10^{-3}$	1949.7
$0.4 \cdot 10^{-3}$	2215.57

---

[3] For details on the Weibull distribution see Appendix B.2.5

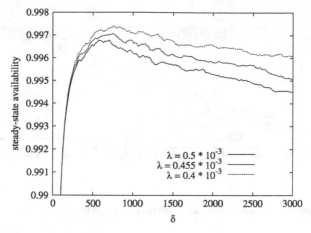

Fig. 7.11 Simulation results for different scale parameters in the failure time distribution

7.5 Stochastic Processes for Shock and Inspection-Based Modelling

In this section we will report on two very elegant mathematical models, a shock model and an inspection model.

In preventive maintenance commonly inspections are used to observe the system state and based on the observation of system degradation preventive maintenance actions are carried out. In [14] two models are investigated for software rejuvenation that use mechanisms known from preventive maintenance. The one uses inspection intervals and an alert threshold rejuvenation policy, the other one is a shock model applying a risk-level rejuvenation policy. Both models use the unavailability due to rejuvenation and crash as the metric of interest which is to be minimised. In the inspection model the alert threshold is the parameter to be tuned, while in the shock model the length of the rejuvenation interval is to be chosen such as to minimise unavailability of a system.

Let us first describe the degradation process, which is the same for both models. We assume a parameter exists that describes degradation and that can be monitored, such as disk usage in a computer system. Let $s(t)$ be the value of the degradation index at time t. Hence, $s(t), t \geq 0$ is the stochastic process modelling system degradation over time.

Let us assume the degradation process is determined by a sum of random shocks, as depicted by the dashed lines in Fig. 7.12. Each shock increases the degradation index by a random variable $X \geq 0$ distributed according to a probability distribution function $F_X(x)$. The distribution of the degradation shocks must be known.

The system is operational as long as $s_{\min} \leq s(t) < s_{\max}$. At $s_{\min}$ the system is fully operational and $s_{\max}$ is the maximum degradation, at which the system will crash. Let T be the random variable denoting the time until $s(t)$ reaches $s_{\max}$ for the first time, i.e.

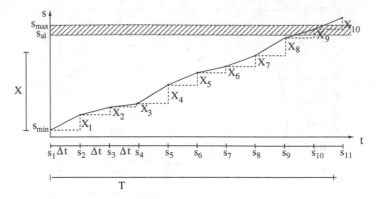

Fig. 7.12 Equidistant observation intervals to determine an alert threshold

$$T = \min\{t : s(t) \geq s_{\max}\}. \tag{7.21}$$

T is the time until the system crashes, distributed according to a cdf $D_T(t, s_{\max}) = Pr\{T \leq t\}$. The survival function is $\bar{D}_T(t, s_{\max}) = 1 - D_T(t, s_{\max})$.

In the following we will study the inspection interval approach and the shock approach. The main difference between the two being that in the inspection interval approach time intervals are given and the value of the degradation index is sought and an alert value in space is determined using the model. In the shock model knowledge of the degradation process is assumed and the rejuvenation interval is sought. The two models are complementary in that one assumes knowledge on timing and computes values in space, while the other one assumes knowledge in space and determines the timing. In the following subsections both models are studied in more detail.

7.5.1 The Inspection Model with Alert Threshold Policy

Let us first consider the inspection model with alert threshold policy. In this model the degradation index is monitored repeatedly and as it hits an alert threshold rejuvenation is triggered. Starting from observation times, the degradation index $s(t)$ is monitored at equispaced intervals of length Δt, as shown in Fig. 7.12. This gives a sequence of degradation values $s(k \cdot \Delta t), k = 0, 1, \ldots$. Let X_k be the increment in the degradation index at time k, then the cumulative degradation can be expressed as the random variable h_k, where

$$h_k = s(k \Delta t) = s_{\min} + \sum_{i=1}^{k} X_i. \tag{7.22}$$

The increment is a strictly positive random variable and therefore the cumulative degradation is monotonically increasing. This allows us to express the survival function as

$$\bar{D}_T(k\Delta t) = Pr\{h_k < s_{max}\}. \qquad (7.23)$$

If the increments are iid then

$$\bar{D}_T(k\Delta t) = \int_0^{s_{max}-s_{min}} f_X^{*k}(z)dz, \qquad (7.24)$$

where $f_X^{*k}(z)dz$ is the k−fold convolution of the density $f_X(x)$ corresponding to the cdf $F_X(x)$.

A treatment of practical issues for observation of software degradation levels in practice can be found in [133].

7.5.1.1 Alert Threshold Rejuvenation Policy

To avoid a system crash, the system needs to be restarted before an inspection determines that the degradation index has crossed an acceptable limit and hence the system has crashed already. Therefore, a warning is needed before the degradation index reaches s_{max}, so that the system can be rejuvenated in time. The *alert threshold policy* introduces this alert value of the degradation index. If at inspection the degradation index exceeds the alert threshold rejuvenation is initiated. Proper choice of the alert value is an optimisation problem. If the alert value is too low, rejuvenation will be carried out too often, leading to unnecessary unavailability of the system. But if the alert threshold is chosen too high, there is not enough time for rejuvenation to prevent the system from crashing.

We introduce a threshold $s_{al} \leq s_{max}$ and as soon as the system degradation reaches that value, rejuvenation is performed. In other words, if

$$s((k-1)\Delta t) \leq s_{al} \text{ and } s(k\Delta t) > s_{al}$$

i.e. the system jumps above the alert threshold in the k-th inspection interval, then rejuvenation is performed. There are two possible situations

1) $s_{al} < s(k\Delta t) < s_{max}$
2) $s(k\Delta t) > s_{max}$.

The distance of s_{al} and s_{max} determines the probability of both events. The appropriate absolute distance $|s_{max} - s_{al}|$ or the relative distance

$$\sigma_{al} = \frac{s_{al}}{s_{max}}$$

both maximise the probability of the first event.

To precisely formulate the probabilities of rejuvenation or crash at the k-th inspection interval we need the random variable T, denoting the first passage time when the degradation index $s(t)$ crosses a value s in the k-th inspection interval for the first time. Let $D_T(k\Delta t, s)$ denote the corresponding cdf and $d_T(k\Delta t, s)$ its probability density. Then $d_T(k\Delta t, s)ds$ is the probability that the degradation index s achieves a value in the interval $[s, s + ds]$ in the k−th inspection interval.

For the alert threshold s_{al} the probability that rejuvenation is performed in the k−th interval is computed as

$$\Pr\{\text{rejuvenate at } k\text{-th step}, s_{al}\} = \int_{s_{min}}^{s_{al}} d_T((k-1)\Delta t, z)(F_{X_k}(s_{max} - z) - F_{X_k}(s_{al} - z))\, dz$$

$$(7.25)$$

and the probability that the system crashes in the k−th inspection interval equals

$$\Pr\{\text{crash at} k\text{-th step}, s_{al}\} = \int_{s_{min}}^{s_{al}} d_T((k-1)\Delta t, z)(1 - F_{X_k}(s_{max} - z))\, dz. \quad (7.26)$$

If the increments of the degradation index are iid random variables with cdf $F_X(x)$, then $F_{X_k}(x) = F_X(x)$ and $d_T(k\Delta t, s) = f_X^{*k}(s)$ and we can simplify (7.25) and (7.26) as

$$\Pr\{\text{rejuvenate at } k\text{-th step}, s_{al}\} = \int_{s_{min}}^{s_{al}} f_X^{*(k-1)}(z)(F_X(s_{max} - z) - F_X(s_{al} - z))\, dz$$

$$(7.27)$$

and

$$\Pr\{\text{crash at } k\text{-th step}, s_{al}\} = \int_{s_{min}}^{s_{al}} f_X^{*(k-1)}(z)(1 - F_X(s_{max} - z))\, dz. \quad (7.28)$$

7.5.1.2   Reward Measures

In order to evaluate the alert policy a metric of interest must be defined. We want to define a reward that represents the cost due to rejuvenation or repair and that is associated with the renewal process of times between successive rejuvenation or repair actions. Let the interrenewal time Y_n be the time interval between the repair or rejuvenation actions. It corresponds to the first passage time of the degradation index $s(t)$ across the alert threshold s_{al}. Irrespective of whether rejuvenation or repair is triggered, the system is renewed when the degradation index crosses the alert threshold.

The expected value of the interrenewal time Y_n equals

$$\tau = E[Y_n] = \sum_{k=1}^{\infty} k\Delta t \cdot \Pr\{\text{crossing } s_{\text{al}} \text{ at the k-th step}\} \tag{7.29}$$

$$= \sum_{k=1}^{\infty} k\Delta t \cdot \int_{z=0}^{s_{\text{al}}} d_T((k-1)\Delta t, z)(1 - F_{X_k}(s_{\text{al}} - z)) \, dz. \tag{7.30}$$

The probability of crossing the alert threshold can be evaluated as

$$\Pr\{\text{crossing } s_{\text{al}} \text{at the k-th step}\} \doteq \bar{D}_T((k\text{-}1)\Delta t, s_{\text{al}}) - \bar{D}_T(k\Delta t, s_{\text{al}}) \tag{7.31}$$

and therefore (7.30) simplifies to

$$\tau = \sum_{k=1}^{\infty} k\Delta t (\bar{D}_T((k\text{-}1)\Delta t, s_{\text{al}}) - \bar{D}_T(k\Delta t, s_{\text{al}})) = \Delta t \sum_{k=0}^{\infty} \bar{D}_T(k\Delta t, s_{\text{al}}). \tag{7.32}$$

At the end of the n−th renewal period a reward R_n is paid (or earned). We assume c_1 as rejuvenation costs and $c_1 + c_2$ to be the cost of repair if a crash happens before the rejuvenation action. c_2 is the extra cost for repair, as compared with the rejuvenation costs. The costs are the same in all renewal cycles. The cost of each renewal is

$$R_n = \begin{cases} c_1 & \text{if } s_{\text{al}} < s(k\Delta t) < s_{\max} \\ c_1 + c_2 & \text{if } s(k\Delta t) > s_{\max} \end{cases} \tag{7.33}$$

for some $k \in 1, \ldots, n$. The total accumulated cost until time t is then

$$C(t) = \begin{cases} 0 & \text{if } N(t) = 0 \\ \sum_{n=1}^{N(t)} R_n & \text{if } N(t) > 0 \end{cases} \tag{7.34}$$

where $N(t)$ is the stochastic process counting renewal events. The process $\{C(t), t \geq 0\}$ is a renewal reward process [87]. The average cost paid (or reward earned) at each cycle evaluates to

$$r = E[R_n] = c_1 \cdot \Pr\{\text{rejuvenate, } s_{\text{al}}\} + (c_1 + c_2)\Pr\{\text{crash at, } s_{\text{al}}\} \tag{7.35}$$

where

$$\Pr\{\text{rejuvenate}, s_{\text{al}}\} = \sum_{k=1}^{\infty} \Pr\{\text{rejuvenate at } k\text{-th step}, s_{\text{al}}\} \tag{7.36}$$

$$\Pr\{\text{crash}, s_{\text{al}}\} = \sum_{k=1}^{\infty} \Pr\{\text{crash at } k\text{-th step}, s_{\text{al}}\}, \tag{7.37}$$

and

$$\Pr\{\text{rejuvenate}, s_{\text{al}}\} + \Pr\{\text{crash}, s_{\text{al}}\} = 1.$$

Assuming the increments of the degradation index to be iid random variables with cdf $F_X(x)$ the probabilities $\Pr\{\text{rejuvenate at } k - \text{th step}, s_{\text{al}}\}$ and $\Pr\{\text{crash at } k - \text{th step}, s_{\text{al}}\}$ were given in (7.25) and (7.26) or in (7.27) and (7.28).

The long-term cost rate can be expressed using a result from the theory of renewal processes. If $r < \infty$ and $\tau < \infty$ then

$$\lim_{t \to \infty} \frac{C(t)}{t} = \frac{r}{\tau} \quad \text{with probability 1.} \tag{7.38}$$

For the inspection model under the alert policy the long-term cost rate is obtained by substituting (7.30) and (7.35) into (7.38) resulting in

$$\lim_{t \to \infty} \frac{C(t)}{t} = \frac{c_1 \cdot \sum_{k=1}^{\infty} \int_{s_{\text{min}}}^{s_{\text{al}}} f_X^{*(k-1)}(z)(F_X(s_{\text{max}} - z) - F_X(s_{\text{al}} - z))\, dz}{\Delta t \sum_{k=0}^{\infty} \bar{D}_T(k\Delta t, s_{\text{al}})} \tag{7.39}$$

$$+ \frac{(c_1 + c_2) \cdot \int_{s_{\text{min}}}^{s_{\text{al}}} f_X^{*(k-1)}(z)(1 - F_X(s_{\text{max}} - z))\, dz}{\Delta t \sum_{k=0}^{\infty} \bar{D}_T(k\Delta t, s_{\text{al}})}$$

again assuming the simplified case where the increments of the degradation index are iid random variables with cdf $F_X(x)$.

The long-term cost rate is plotted in [14] for the following parameter values:

$F_X(x)$ is the Gamma distribution with three different sets of parameters, as to achieve identical expected value, but different variances.

- $F_X(x) = \Gamma(0.2, 0.2)$. The cdf of the increment is decreasing hazard rate (DFR) with expected value equal to 1 and variance 5. The coefficient of variation equals 2.24.
- $F_X(x) = \Gamma(1, 1)$. The cdf of the increment exponentially distributed (constant hazard rate) with expected value, variance and coefficient of variation all equal to 1.
- $F_X(x) = \Gamma(5, 5)$. The cdf of the increment is an Erlang-5 (IFR - increasing hazard rate) distribution with expected value equal to 1, variance equal to 0.2 and coefficient of variation equal to 0.447.

Furthermore, the chosen failure profile is

- The feasible values of the degradation index are in the interval $[0, 10)$, i.e. $s_{min} = 0$ and $s_{max} = 10$.
- The downtime to recover from a rejuvenation is $c_1 = 15$ min, the downtime to recover from a crash is $c_1 + c_2 = 45$ min.
- The degradation index is monitored once per day ($\Delta t = 1$ day).

In [14] in addition realistic data from a data base system has been studied.
An inspection-based model with alert threshold has been used for a closed loop design in [71] demonstrated by rejuvenating a web server. The degradation model corresponds to the inspection model in Sect. 7.6. The alert threshold policy is compared with executing no rejuvenation and a fixed rejuvenation interval. It has been shown in [71] that especially in random environments where degradation is not deterministic adaptation of the rejuvenation interval is beneficial, even though it incurs a higher cost than just employing a fixed size interval. The feed-back rejuvenation methods have been included in IBM's Director tool for performance management of the xSeries servers [21].

7.5.2 The Shock Model with a Risk Policy

Instead of making assumptions on the observation of degradation, we now look at the degradation process itself and assume degradation to take place due to random shocks. The random shocks not only determine the degree of degradation coming along with each shock, i.e. the 'height' of the shock, but also the frequency of occurrence, or the time between shocks. The time instants at which the shocks happen give rise to a point process $S(t)$. This point process determines the number of shocks $S(t)$ in the interval $[0, t)$. The increment of the degradation index at each shock again is a random variable X with cdf $F_X(x)$. To specify the degradation process we first need to know the probability $P_k(t)$ of seeing k shocks until time t, i.e.

$$P_k(t) = \Pr\{S(t) = k\}$$

which is the second term in the survival function

$$\bar{D}_T(t) = \sum_{k=0}^{\infty} \bar{D}_T(t|k) \cdot \Pr\{S(t) = k\}. \tag{7.40}$$

$\bar{D}_T(t|k)$ is the survival probability at time t, under the condition of having had k shocks in the time interval $[0, t)$. As in the inspection model, we make the simplifying assumption that the increment of the degradation process is a random variable X with cdf $F_X(x)$ and density $f_X(x)$, we assume that shocks are independent and the degradation process is additive (Fig. 7.13). Then $\bar{D}_T(t|k)$ can be expressed as (see (7.24))

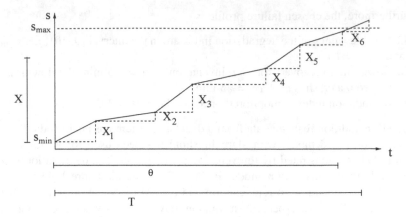

Fig. 7.13 Degradation process to determine rejuvenation interval

$$\bar{D}_T(t|k) = \int_0^{s_{max}-s_{min}} f_X^{*k}(z)dz, \tag{7.41}$$

Until now we have described the height of the increment of the degradation index with each shock. To fully specify the degradation process we still need a stochastic process describing the number of shocks in a time interval, or the time between shocks. In degradation processes it is common to model the aging of a system with a time-dependent shock rate (or time between shocks). The point process $S(t)$ is a non-homogeneous Poisson process and a power-law dependence for the time-dependent rate $\lambda(t)$ gives

$$\lambda(t) = c\beta t^{\beta-1}, \tag{7.42}$$

where for $\beta > 1$ the interarrival time between shocks is decreasing with time (age of the system) and the shock rate increases, for $\beta < 1$ vice versa. If $\beta = 1$ then $S(t)$ is a Poisson process. Using the time-dependent shock rate we can write the probability of having k shocks until time t as

$$P_k(t) = e^{-ct^\beta} \frac{(ct^\beta)^k}{k!} \tag{7.43}$$

and the survival probability $\bar{D}_T(t)$ as

$$\bar{D}_T(t) = \sum_{k=0}^{\infty} \bar{D}_T(t|k) \cdot e^{-ct^\beta} \frac{(ct^\beta)^k}{k!}. \tag{7.44}$$

In [14] earlier work is pointed out, deriving general and interesting results for the Poisson distributed random shock model. If only the increment of the degradation index X is a positive random variable, the survival probability $\bar{D}_T(t)$ is increasing hazard (failure) rate on the average. Hence, independent of the distribution of the single shocks, the overall process on the average represents degradation.

7.5.2.1 Risk Level Rejuvenation Policy

After having described the degradation process in the shock model, we now proceed to formulating a rejuvenation policy. Given the probability distribution function $D_T(t, s_{\max})$ denoting the probability that the degradation index will cross the threshold $s_{\max}$ and crash until time t the rejuvenation interval θ is chosen such that the probability of a crash stays below a threshold, or risk level $(1 - \alpha)$. At a confidence level α (i.e. $\alpha = 0.95$) the system is rejuvenated at time $\theta(\alpha)$ where

$$D_T(\theta) = 1 - \alpha. \tag{7.45}$$

The value θ is the $(1 - \alpha)$ percentile of the distribution $D_T(t)$ and, obviously, the higher the confidence α the shorter becomes the rejuvenation interval θ. Hence θ depends on the choice of α and we may express this in the subscript θ_α.

7.5.2.2 Reward Measures

The formulation of the reward measures more or less follows along the same lines as in the inspection-based model with alert threshold. We describe the renewal process Y_n, consisting in the periods between successive crashes or rejuvenations and associate costs with both crash and rejuvenation. The accumulated reward then is the accumulated cost of crash and rejuvenation and this corresponds to the unavailability of the system. Unavailability of the system depends on the length of the rejuvenation interval θ, which again depends on the risk level α.

Let us assume that crashes and rejuvenation times form renewal times and that T_n as defined in (7.21) is the time until the system crashes in the n-th cycle. The duration of the n-th renewal cycle is defined as

$$Y_n = \min(T_n, \theta_\alpha).$$

The reward paid (or earned) at the end of the n-th renewal cycle is similar to the reward in the inspection model with alert threshold (7.33)

$$R_n = \begin{cases} c_1 & \text{if } Y_n = \theta_\alpha \\ c_1 + c_2 & \text{if } Y_n = T_n < \theta_\alpha. \end{cases} \tag{7.46}$$

The total cost $C(t)$ is defined as in (7.34) and $r = \mathrm{E}[R_n]$ again is the expected value of the reward R_n gained at the end of each renewal cycle. The expected interrenewal time is again $\tau = \mathrm{E}[Y_n]$. Both are now defined as

$$\tau = \mathrm{E}\,[Y_n] = E(\min(T_n, \theta_\alpha)) = \int_0^{\theta_\alpha} (1 - D_T(t))\, dt = \int_0^{\theta_\alpha} \bar{D}_T(t)\, dt \quad (7.47)$$

$$r = \mathrm{E}\,[R_n] = c_1 + c_2 D_T(\theta_\alpha) = c_1 + c_2(1 - \alpha). \tag{7.48}$$

The long-term cost rate is computed using (7.38), which now evaluates to

$$\lim_{t \to \infty} \frac{C(t)}{t} = \frac{c_1 + c_2(1 - \alpha)}{\displaystyle\int_0^{\theta_\alpha} \bar{D}_T(t)\, dt}. \tag{7.49}$$

To obtain a low risk of system crashes $(1 - \alpha)$, α should be chosen close to 1. This leads to very short rejuvenation intervals and hence a considerable unavailability due to rejuvenation. Clearly, wanting to avoid downtime due to system crashes α should be chosen large, while to reduce unavailability due to rejuvenation, α should be chosen small. Considering the downtime costs incurred by crashes and rejuvenation, respectively, there will be an optimal value of α (and hence of the rejuvenation interval θ) that minimises the overall cost.

7.6 Inspection-Based Modelling using the Möbius Modelling Tool

The model in [158] takes a slightly different approach on inspection-based preventive maintenance as it models the system in more detail. The model allows for system crashes and repair during inspection intervals and uses different degrees of maintenance depending on the degradation level. In [158] a Markov regenerative process with a subordinated semi-Markov reward process is defined using an arbitrary number of inspection intervals and allowing for arbitrary probability distributions for the inspection interval as well as for the inspection activity, the repair after a component as well as a full system failure. The computed metrics are the expected downtime and the expected cost over an interval $[0, T]$ where $T = 1{,}000$ h. The model is solved using deterministic as well as exponentially distributed inspection intervals δ, while all other parameters are assumed exponentially distributed.

We use the Möbius modelling tool [31] to model and simulate an instance of the model as solved analytically in [158]. Figure 7.14 shows a stochastic activity network as atomic model in Möbius with only $n = 4$ stages of deterioration. The general model in [158] is depicted as a state transition diagram.

The model in Fig. 7.14 initially has a token in place $D0$, the new and up state. The state of the system deteriorates with the firing of transitions $t0$ through $t3$ until it finally fails. If the firing time of transitions $t0, .., t3$ is exponentially distributed then the failure time distribution is a hypo-exponential distribution, which has an increasing hazard rate and therefore models aging.

At each stage of deterioration the system may fail and can be restarted. This is modelled by the transitions tp_i and tr_i. Inspection is initiated upon the firing of td_i and is completed when $tins_i$ fires. If the system has reached deterioration

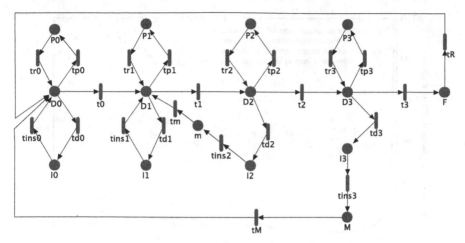

Fig. 7.14 Atomic Möbius model for inspection-based preventive maintenance

grade $D2$ inspection is followed by a minimal maintenance (firing of transition tm), which improves the system state such that it returns to deterioration state $D1$. Once the system has reached deterioration state $D3$ a major maintenance is needed after which the system is 'as good as new' (back in state $D0$). The failed system needs a full reboot, modelled as the firing of transition tR. All parameter values are taken from [158] and summarised in Table 7.3.

As in [158] we compute the expected downtime and the expected cost over an interval of 1,000 h. The following figure shows that an optimal inspection interval exists for both considered metrics. The length of the optimal inspection interval, however, depends on the chosen metric.

We find that the expected downtime is minimal for an inspection interval of approximately 3.5, while for the expected cost an inspection interval of 1.2 is optimal. Unlike the observation in [158], we find that the optimal inspection interval is irrespective of the distribution of the inspection interval, where we considered

Table 7.3 Parameters of the inspection-based SAN

Parameter	Transition	Description	Value
λ	t_i	Deterioration failure rate	0.1/h
λ_p	tp_i	Poisson failure rate	0.05/h
$1/\mu_r$	tr_i	Mean time for retry	2 min
$1/\mu_R$	tR	Mean time for full reboot	1 h
$1/\mu_m$	tm	Mean time for minimal maintenance	5 min
$1/\mu_M$	tM	Mean time for major maintenance	15 min
$1/\mu_{ins}$	$tins_j$	Mean time for inspection	40 s
c_r		Cost of retry	$5,00/h
c_R		Cost of full reboot	$5,000/h
c_m		Cost of minimal maintenance	$1,500/h
c_M		Cost of major maintenance	$3,000/h
c_{ins}		Cost of inspection	$1,000/h

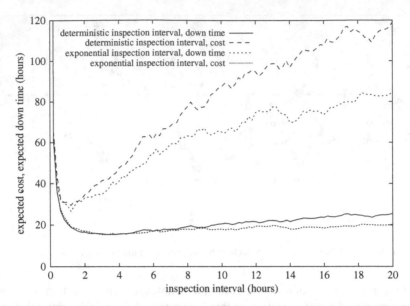

Fig. 7.15 Expected downtime and expected cost for deterministic and exponentially distributed inspection intervals

the exponential distribution and deterministic inspection intervals. The distribution of the inspection interval does influence the results in that for longer deterministic inspection intervals the expected downtime and even more so the expected cost are higher than both would be when using an exponentially distributed inspection interval.

Figure 7.15 furthermore shows that selecting the optimal inspection interval length such that the expected downtime is minimised penalises the expected cost roughly to the same extent as choosing the optimal inspection interval with respect to the expected cost would increase the expected downtime. As we have seen earlier, also in this model when in doubt the inspection interval should be chosen rather too long than too short.

7.7 Comparative Summary of the Stochastic Models

The models in this chapter can be compared with respect to the considered failure and repair characteristics, the metric of interest, the parameter that is tuned, the model class used to formulate the model and the solution method.

The first stochastic model in Sect. 7.1 is historically the first rejuvenation model that was published and deserves credits on its own rights. It corresponds to a CTMC, which means that failure and repair both happen in exponentially distributed time. In consequence, the model does not include aging and therefore in many cases immediate rejuvenation or no rejuvenation achieves the minimal downtime or down-time cost.

The second model, discussed in Sect. 7.2 shows a first attempt to incorporate aging through a time-dependent failure rate. The repair time follows some general distribution and therefore the model can no longer be mapped on a CTMC. This model, as the one in the following section includes a queueing model to represent the processing of tasks. Metrics of interest are system availability as well as loss probability or number of lost jobs. The model in Sect. 7.2 investigates two rejuvenation strategies where the length of the rejuvenation interval is the parameter that is varied. Rejuvenation can either be triggered in regular intervals, or after additionally waiting for the queue to be empty. The model in Sect. 7.3 does not use the different rejuvenation policies. Instead, rejuvenation is performed in deterministic intervals, giving raise to a deterministic and stochastic Petri net (DSPN) model, with exponentially distributed repair time and number of lost jobs as the considered metric.

The model in Sect. 7.4 does not explicitly include the processed jobs. It is represented as a stochastic activity network using Möbius. It includes aging as the failure time is Weibull distributed with increasing failure rate. The rejuvenation interval is sought as to optimise the steady-state availability of the system.

In brief the characteristics of the models are listed in Table 7.4.

The last two models in this chapter take a different point of view and model the degradation process. The first one, presented in Sect. 7.5, uses an analytical formulation and minimises unavailability as well as rejuvenation and repair cost by choosing the appropriate inspection interval. The model is inherently different from all other models in this chapter and is therefore not included for comparison in the table.

The last model in Sect. 7.6, the second model of the degradation process, is again formulated as a stochastic activity network. It incorporates aging through the phases of the hypo-exponential distribution accumulated in the path through the degradation levels. The expected downtime and expected cost are minimised by a good choice of the inspection interval.

The numerical results of the models cannot easily be compared. The model parameters and the model formulation differ in many details. Therefore, the models and their results are not directly comparable and the effects of more realistic models

Table 7.4 Comparison of the stochastic models for rejuvenation

Section	Failure distr.	Repair distr.	Metric	Parameter
7.1	Exponential, no aging	exp.	Downtime, downtime cost	Rejuvenation interval
7.2	Time dependent exp., aging	generally distr.	Availability, loss probability	Rejuvenation strategy + interval
7.3	Exponential, no aging	exp.	No. of lost jobs	Det. rejuvenation interval
7.4	Aging, Weibull IFR	exp.	Steady-state availability	Rejuvenation interval
7.6	Hypo-exponential, aging	exp.	Exp. downtime, exp. downtime cost	(det, exp) inspection interval

cannot be precisely identified. But the structural behaviour of the metric of interest for varying values of the rejuvenation interval, or the inspection interval in most cases show a clear minimum that allows to select the best interval.

7.8 Further Reading

A study that combines several of the concepts used in this chapter is the performability-oriented rejuvenation in [144] that uses accomplishment levels from the performability concept [143] for a distributed system with redundant replicas. A SAN model is used to determine the performability of different rejuvenation policies.

The cost of rejuvenation is closely related to the time penalty of the system being out of operation during rejuvenation. This issue is tackled in [139, 151] by using virtualisation, such that during system restart a different virtual machine takes over processing of the running tasks. The approach has been evaluated using web services benchmarks.

The failure characteristics of a systems are usually unknown. As the failure distribution is needed in most models to determine a suitable rejuvenation frequency assumptions on this distribution are made. A predictive technique is used in [131] to determine bounds on the system availability.

Part IV
Checkpointing

Introduction

In this part we want to briefly recall checkpointing systems and survey existing stochastic models for checkpointing. The models aim at choosing the checkpointing interval such that a metric of interest is being optimised. The models make assumptions on the considered system, on the considered faults and on locations in the system where checkpoints can possibly be placed. In order to introduce practical issues of checkpointing we will first give a brief overview of existing systems work in checkpointing. A huge body of work in systems science appeared over the past decades and work in this field is still ongoing. We do not try to cover systems checkpointing exhaustively in this chapter, since it is beyond the scope of this work. The systems work is introduced only for the purpose of placing the stochastic modelling work in the right context.

Checkpointing is being applied to transaction processing systems as well as to batch programs. It is being applied in production environments, real-time systems [60] as welll as in distributed message passing systems. With checkpointing the system or program state is saved to permanent storage and if a failure happens after the checkpoint, the system or program does not need to restart over from the beginning but only from the most recent checkpoint. Taking a checkpoint often is called a *save*, while restarting from the most recent checkpoint is called a *rollback*. When checkpointing a program environment, the whole system needs to be rolled back to the state where the checkpoint was taken. When checkpointing a process the rollback restores an earlier state of the process. When checkpointing a transaction processing system also the list of waiting and processed transactions has to be saved and when the state is restored the lost transactions from the audit trail must be replayed.

Chapter 8
Checkpointing Systems

Checkpointing applies to large software systems subject to failures. In the absence of failures the software system continuously serves requests, performs transactions, or executes long-running batch processes. If the execution time of the task and the time at which processing starts is known, then the moment of completion of the task is known as well. If failures can happen the completion of a task severely depends on the underling fault model. The typical fault model employed in checkpointing consists in the assumption that faults are detected immediately as they happen. This implies that only crash-faults are considered and no transient or Byzantine faults that would require fault-detection mechanisms. Some checkpointing models assume that faults are detected only at the end of the software module [152].

8.1 Checkpointing Single-Unit Systems

In this section different techniques are outlined for checkpointing monolithic systems.

Sequential checkpointing. Sequential checkpointing is the basic type of checkpointing. It is concerned with uni-process systems. Without checkpointing upon failure of the system typically all work performed so far is lost and the computation has to start anew. Checkpointing aims at reducing the amount of work that is lost upon failure of the system by intermediately saving the whole state of the system. Saving the system state usually comes with some cost, or overhead. Usually this is not clearly specified. The checkpointing cost can be the time needed for checkpointing. It can also be some cost incurred by system operation. The checkpoint latency is the time needed to establish a checkpoint. A checkpoint is also called *recovery point.* In sequential checkpointing the checkpoint latency is equal to the checkpoint overhead. During the checkpoint save operation the system cannot perform any useful work. Therefore, work is delayed because of checkpointing. If no failure happens checkpointing is seen purely as retarding system operation. If a failure happens during computation, on the other hand, then the system does not have to roll back to the initial state, instead it can roll back to the most recent checkpoint. Checkpointing can then considerably reduce the amount of work that is lost upon failure of the system.

K. Wolter, *Stochastic Models for Fault Tolerance,*
DOI 10.1007/978-3-642-11257-7_8, © Springer-Verlag Berlin Heidelberg 2010

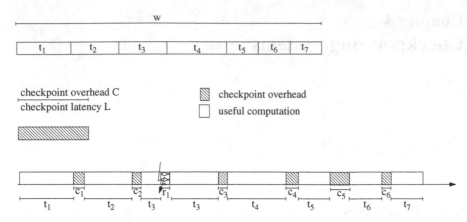

Fig. 8.1 System model

An exemplary checkpointing scenario is depicted in Fig. 8.1, where w denotes the task length, t_i the intervals between checkpoint $i - 1$ and i. The checkpointing cost is c_i and the rollback cost r_i.

There obviously is a trade-off between the amount of work lost when the system fails and the checkpointing overhead. Many stochastic models aim at optimising this trade-off with respect to some performance metric, typically either the expected completion time of the task, or the availability of the software system.

Forked checkpointing When using sequential checkpointing on a uni-processor system no useful work is being performed during the whole save process of a checkpoint. Therefore, in forked checkpointing the process whose state is being saved creates a child process which performs the establishment of the checkpoint while the parent process continues processing of useful work. This reduces the checkpoint overhead when using efficient processors, since very often computation of the parent process can be done in parallel with stable storage access of the child process [156].

The checkpoint latency is no longer equal to the checkpoint cost, as illustrated in Fig. 8.2, where only one checkpoint save operation is depicted.

The probability of a failure during checkpointing increases when spreading out the checkpoint save operation over time. Therefore high latency reduces the performance gain through checkpointing. The quality of forked checkpointing is not evaluated by means of the overhead or the latency, but by the overhead ratio, which is the relative extra time required to perform a job with forked checkpointing [156]. The overhead ratio is formally defined later.

A system crash can terminate all processes, or it can be a partial crash affecting only some processes. An optimistic view assumes that sometimes only the parent process which does the computation is affected and the forked child process can complete the checkpoint save operation [70]. The rollback in these cases is more or less a restart from the current state and very little computation is wasted.

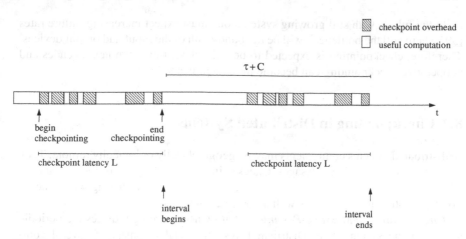

Fig. 8.2 Forked checkpointing

Cooperative checkpointing Cooperative checkpointing is also called *compiler-assisted* checkpointing in the literature [39, 94]. There exist three types of implementations of checkpointing. Checkpointing can be integrated into applications by the application programmer. Checkpointing can also be implemented as a function of the operating system. This is called *system-initiated* checkpointing [135]. Cooperative checkpointing [115] combines both application-initiated as well as system-initiated checkpointing. At runtime the operating system uses heuristics classifying the system state to decide whether a checkpoint, which is implemented in the application code, should be executed.

Most applications use equidistant checkpoints, which, as we will see later in this chapter, is in many cases the best choice. Cooperative checkpointing then appears to be irregular, since the system can either grant or deny a checkpoint requested by the application. For realistic distributions of the time between failures cooperative checkpointing can perform up to four times better than periodic checkpointing [115]. A good heuristic of the system state is essential for the performance of cooperative checkpointing. In [116] two strategies for cooperative checkpointing are investigated. Work-based cooperative checkpointing trades the cost of executing a checkpoint against the risk when skipping it. Risk-based cooperative checkpointing is a slight modification of work-based cooperative checkpointing, where the risk is not measured as the worst case loss if a failure happens, but the expected loss. If a failure can be predicted with 10% accuracy, cooperative checkpointing can considerably reduce the amount of work lost, increase system utilisation and improve other system metrics as compared with not checkpointing at all. If 50% of all failures are known ahead of time, then the cooperative checkpointing strategy performs as well as periodic checkpointing, for over 50% an improvement over periodic checkpointing is possible. Both types of cooperative checkpointing do not differ much in performance [116].

In the future, with still growing systems one must expect increasing failure rates [39] where still the bottleneck will be the bandwidth of the input and output devices. Therefore, checkpointing is expected to be still relevant over the next decades and cooperative checkpointing can become more significant.

8.2 Checkpointing in Distributed Systems

In distributed systems components may be geographically or logically remote. They cooperate by exchanging messages. Checkpointing of such systems poses a number of challenging problems. This section discusses the problems arising when check-pointing distributed systems as well a possible solution.

Checkpointing in Message-Passing Systems Checkpointing requires the periodic saving of the system state. Distributed systems usually consist of several components that can be located geographically and logically apart from each other. To checkpoint a distributed system the state of all components must be saved at the same time. Time synchronisation in distributed system is a difficult issue itself, but it would not be relevant if the components of the system were not to communicate with each other. The exchange of messages turns the check-point save operation into a difficult task. Even if perfect time synchronisation were available this would not guarantee consistent roll back recovery, because messages that were sent before the checkpoint save operation may not yet have arrived. In the sending process the system state includes information on having sent the message while the receiving process does not include receipt of the message in its saved copy of the system state. The basic problem is illustrated in Fig. 8.3.

A *cut* (or *recovery line*) through the system defines the point where the state of each process in the distributed system is saved in a checkpoint. The system in Fig. 8.3 consists of the three processes P_1, P_2 and P_3, which exchange messages. The graph on the left in Fig. 8.3 shows a consistent cut, because in the check-pointed state of the system all messages that have been received, also have been sent. The graph on the right in Fig. 8.3 shows an inconsistent cut, because the saved system state includes received messages that have not been sent at the point

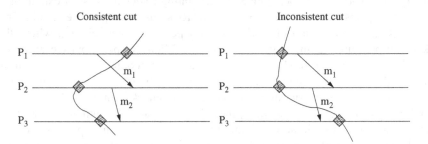

Fig. 8.3 Consistent and inconsistent recovery line

that they were saved in the checkpoint. It should be noticed that even a consistent cut alone does not guarantee consistency of the system at rollback recovery. The sent messages, which have not been received before taking the checkpoint must be redelivered by the checkpointing protocol. This mechanism is similar to the repeated delivery of messages in reliable communication protocols, such as TCP. Messages that have been sent, but not received, at the time a checkpoint is taken, are called *in-transit* messages. In Fig. 8.3 on the left message m_2 is an in-transit message.

To create consistent cuts requires some coordination between the communicating processes. If processes autonomously save checkpoints and an inconsistent cut is created then the system detects this inconsistency at the rollback recovery operation and rolls back to the previous checkpoint. If that checkpoint again is inconsistent the system rolls back to the next preceding checkpoint. This may cause a step wise rollback to the initial state of the system, which is called the *domino effect* and is highly undesirable.

The possible domino effect is the main disadvantage of *uncoordinated* checkpointing, where each process decides individually when to take a checkpoint. Its advantage is that it has low administration costs. *Coordinated* checkpointing on the other hand assumes a central control unit which stops communication between the processes to avoid inconsistency of checkpoints. The third possible way to create a checkpoint is to induce the checkpoint save operation by communication messages between all participating processes. This is a very expensive method and it is not always known what processes will communicate if they have not done so yet. These are the disadvantages of *communication-induced* checkpointing. Strong on the positive side is that handshake protocols are very well studied and have known properties and performance.

When using a general fault model that allows for failures during checkpointing and rollback recovery not only the checkpoint save operation but also the rollback recovery can be carried out in different ways. Two different rollback recovery algorithms are analysed in [148], the sequential and the lookahead rollback. Sequential rollback means that upon failure of the system it is rolled back to the most recent checkpoint and if recovery fails the system is again rolled back to the previous checkpoint. This happens sequentially and potentially until the initial system state is reached. In lookahead rollback a cost function of recovery at the individual checkpoints is defined and all existing checkpoints are considered for selection. The most recent checkpoint whose predecessor has higher recovery cost is then selected. This scheme is bounded by sequential rollback. The models in [148] show that the lookahead rollback achieves lower expected rollback and recovery cost than the sequential rollback, but in the optimum requires more checkpoints. Similarly, an adaptive checkpointing-recovery cost model can be used to evaluate the *convenience* [125] of the checkpoint before it is saved.

An exhaustive survey of checkpointing protocols for distributed systems is given in [40] and a very thorough classification of checkpointing algorithms can be found in [76]. Interesting experience reports of checkpointing in practice are given in [98, 173].

Stochastic models for checkpointing do not consider any problems related to checkpointing in distributed system. Therefore, we do not go in more detail into this huge area of existing work.

For stochastic modelling two types of systems are distinguished: systems processing large batch processes and transaction processing systems. Checkpointing batch programs requires saving a possibly very large system state, while a checkpoint of a transaction processing system consists of two components: the system state at the time the checkpoint is taken and the audit trail (the list of transactions processed after the checkpoint save operation). The audit trail is being reprocessed in the recovery process after the system has rolled back to the most recent checkpoint. For batch programs processing is continued after the rollback under the assumption that this completely recovers the system state at the time of failure.

A particular challenge is the checkpointing of super computers. As they process many large tasks their fault-tolerance is important. Studies exist of checkpointing IBM's Blue Gene/L [116] and Blue Gene/P [110] machine.

Chapter 9
Stochastic Models for Checkpointing

In the 1970s until early 1990s a huge amount of work on modelling checkpointing has been published and we will cover in this section the most important models, insights, algorithms and results. Models of checkpointing differ in the granularity at which the environment is included in the model, if at all. They also differ in which components of the model are considered deterministic and which components are modelled as random variables, and what system characteristics are included at all. Some models allow for checkpoints being taken and possible failures during recovery, some allow for one of the two or none. Some models assume checkpoints to be equidistant, others consider them to be taken at random time intervals. There are many ways to organise and structure existing work in modelling of checkpointing. We will use the structure given in [113], where checkpointing schemes are divided in system level and program level checkpointing. In [113] only program level checkpointing is regarded in detail. We will here summarise existing work in both fields.

A very similar distinction can be found in [23], the first survey article on analytic models for checkpointing, summarising the five previous papers that existed at the time. In [23] program level checkpointing is called checkpointing of a process control system, which repeatedly runs the same long job or a small set of jobs. The execution time of a job is typically very long as compared with the MTTF. The equivalent of system level checkpointing is on the other hand checkpointing of a data base system with many small jobs. The execution time of these jobs typically is short as compared with the MTTF of the system. A combination of the recovery block concept and checkpointing is used to derive a model in [77].

The purpose in program level checkpointing is to minimise the checkpointing overhead, by minimising rollback and recovery time. As a consequence the job execution time is minimised. In a stochastic model the cost of rollback and recovery due to failures has to be traded against the cost of taking checkpoints. Usually an optimal checkpointing interval can be determined.

The purpose of system level checkpointing, on the other hand, is to optimise availability of a computing system. This increases fault tolerance of the system. In transaction processing systems (in [23] database systems are considered) taking a checkpoint involves storing the system state and keeping an audit trail of transactions carried out since the checkpoint. Storing the system state usually is a costly and

K. Wolter, *Stochastic Models for Fault Tolerance*,
DOI 10.1007/978-3-642-11257-7_9, © Springer-Verlag Berlin Heidelberg 2010

time consuming task. Stochastic models of those systems aim at tuning the check-pointing interval so as to optimise system availability with respect to the trade-off between the cost of checkpointing and the cost of rollback and replay of the audit trail.

When comparing the two types of checkpointing first the metric of interest is slightly different. In program level checkpointing the completion time of a job is considered, whereas in system level checkpointing the metric of interest is the avail-ability of the system. Checkpointing is just a saving of the complete system state in program level checkpointing, whereas in system checkpointing it also requires logging an audit trail. Typically, in program level checkpointing the system is mod-elled using a simple model with two discrete states, in system level checkpointing sometimes more complex system models are used. Assumptions on failures and distribution of failures are typically more complex in program level checkpointing, since this sometimes is considered with real-time control jobs or other critical tasks [23]. Some models cannot clearly be grouped into one of these two types. In [90] a system level checkpointing model is considered, but tasks take very long, which is typically not the case in transaction processing systems. If tasks take very long, the system spends a long time processing one task and even though the model is a queueing model with different server states the analysis is concerned with the completion of single tasks, as it is common in program level checkpointing. So the model has characteristics of both, system level and program level checkpointing.

9.1 Checkpointing at Program Level

In checkpointing at program level we consider systems processing one (or possi-bly more) long-running jobs. The job length in a failure free system is known in advance [93, 51, 113]. System failure increases the time needed to process the given amount of work. The models for the completion time of tasks in unreliable systems discussed in the first part of this book can be used as reference models in order to judge whether checkpointing is beneficial in such a systems. In this chapter we discuss stochastic models for the completion time of tasks in unreliable systems that include checkpointing.

The completion time of a job is the time from starting the job until its completion. Completion time includes the pure processing of the job, but also the time needed to take a checkpoint as well as the reprocessing time after a failure.

We have seen in Chap. 2 that in a system subject to failures, where with a fail-ure the affected job needs to be restarted from beginning the completion time of a job increases exponentially with the work requirement. In this chapter we will demonstrate how checkpointing transforms this increase to linear growth, i.e. when applying checkpointing the time needed to finish a certain amount of work increases only linearly with the amount of work [37, 90].

The easiest and most straightforward way to do checkpointing would be to divide the task length into $k > 1$ equally long intervals and save the system state at the end of each of the first $k - 1$ intervals. At the end of the last interval the job is completed and no checkpoint needs to be taken [30, 113]. This strategy assumes that checkpoints can be placed at arbitrary points in the program code and that the processing time for parts of code is known in advance and identical with every program run. These assumptions generate an abstraction and a simplification that might not always be appropriate. More complex and perhaps more realistic is the random placement of checkpoints, where the intervals between checkpoints are random variables distributed according to some probability distribution.

As in [113] we will in this section distinguish the two cases of equidistant checkpointing and checkpointing at random intervals. The purpose of checkpointing typically is to reduce the job completion time by optimising the time between checkpoints. But also the probability of task completion is computed.

More recent work [70] takes a very different angle and aims at reducing the recovery time instead of optimising the checkpointing effort. The analysis in [70] is applied to forked checkpointing. Minimising the checkpoint overhead ratio in forked checkpointing is addressed in [156].

9.1.1 Equidistant Checkpointing

The system model used to determine job completion time with checkpointing is the same as the one in Chap. 2 for systems without checkpointing. The system model is depicted in Fig. 2.1 on p. 13 and all parameters are identical to those in Chap. 2. The time between failures and the time to repair are again assumed exponentially distributed and hence the hazard rate functions $h_U(t)$ and $h_D(t)$ evaluate to γ and ν respectively. In models for checkpointing instead of the time to repair the time to *rollback* to the previous checkpoint is considered. Rollback is an important concept in checkpointing. It includes the time to restart the system and restore the state at the most recent checkpoint. The model in this section does not consider the actual repair of the system explicitly. We will use D to denote constant rollback time.

The processing requirement w is divided into k equal parts and there is a checkpoint taken at the end of each part, except for the last one, hence there are in total $k - 1$ checkpoints taken. Checkpoint duration is assumed to be a random variable C with CDF $F_C(t)$, pdf $f_C(t)$ and LST $F_{\tilde{C}}(s) = \int_0^\infty e^{-st} \, dF_C(t)$ time units.

The total execution time needed to perform the work requirement is a random variable denoted $T(w, k)$ with CDF $F_T(w, k)$. The LST of the distribution of the job completion time with checkpointing and failures is given and proven in [113] as

$$
F_{\tilde{T}}(s, w, k) = \left(\frac{(s + \gamma) F_{\tilde{C}}(s + \gamma) e^{-(s+\gamma)w/k}}{s + \gamma(1 - F_{\tilde{D}}(s)(1 - F_{\tilde{C}}(s + \gamma) e^{-(s+\gamma)w/k}))} \right)^{k-1} \times
$$

$$
\left(\frac{(s + \gamma) e^{-(s+\gamma)w/k}}{s + \gamma(1 - F_{\tilde{D}}(s)(1 - e^{-(s+\gamma)w/k}))} \right) \quad (9.1)
$$

Similar as for (2.1) on p. 14 also for (9.1) no solutions are available in the literature, indicating that the inverse transformation must be a challenge. But there exists a solution for the expected job completion time with checkpointing in the presence of failures [113]:

$$E\{T(w,k)\} = \left(\frac{1}{\gamma} + E\{D\}\right)\left((k-1)(F_{\widetilde{C}}(-\gamma)e^{\gamma w/k} - 1) + (e^{\gamma w/k} - 1)\right).$$

(9.2)

Not even the formula for the expected completion time provides us with an expression that can be evaluated directly. This explains why LST expressions, although they form a mathematical model, are not very popular in this context and rarely find their way into implementations in practice.

Following [113], if we assume that $E\{T(w,k)\}$ is a non-concave function in k, then equidistant checkpointing is beneficial only if the expected task completion time using one checkpoint is less than the expected task completion time without checkpointing, i.e. $E\{T(w,2)\} < E\{T(w,1)\} = E\{T(w)\}$, which implies $e^{\gamma x/2}(F_{\widetilde{C}}(-\gamma) + 1 - e^{\gamma x/2}) < 1$. The convexity issue for the completion time distribution considering different probability distributions of the pieces of work performed between checkpoints is discussed in [30]. Since C is a non-negative random variable $F_{\widetilde{C}}(-\gamma) \geq 1$ and for $w \to 0$ the above inequality does not hold checkpointing is not beneficial if the work requirement is very small. This corresponds to our intuition.

Expressing the work requirement w as k multiples of a fixed program length τ, $w = k \cdot \tau$ and correspondingly $k = w/\tau$ allows to reformulate (9.2) as

$$E\{T(w,w/\tau)\} = \left(\frac{1}{\gamma} + E\{D\}\right)\left(\left(\frac{w}{\tau} - 1\right)(F_{\widetilde{C}}(-\gamma)\, e^{\gamma\tau} - 1) + (e^{\gamma\tau} - 1)\right).$$

(9.3)

Since τ in (9.3) is a constant that does not change as w increases (assuming that w/τ will still be an integer) (9.3) shows a linear increase of the expected completion time as the processing requirement increases. This is in contrast to the exponential increase in (2.2) shown in Fig. 2.2 and discussed in [90].

In [113] it is further shown that the optimal checkpointing interval τ^* does not depend on the job length. Instead, it solely depends on the system failure rate and the distribution of the time needed to checkpoint the system, leaving recovery time an open degree of freedom. But no closed-form expression or algorithm for computing the optimal checkpointing interval is given. As we will see now, for exponentially distributed times between failures a closed-form solution as well as useful approximations exist.

A very simple first order approximation is given in [181]. For exponentially distributed time between failures, deterministic checkpointing cost C and constant time intervals between checkpoints τ the time lost due to checkpointing and failures τ_l can be expressed as

$$\tau_l = \frac{1}{\gamma} + \frac{C}{1 - e^{\gamma(C+\tau)}}. \tag{9.4}$$

The optimal time interval between checkpoints that minimises the amount of time lost τ_l is obtained by differentiating τ_l with respect to τ and equating the derivative to zero.

$$\frac{d\tau_l}{d\tau} = \frac{1 - e^{\gamma(\tau+C)} - \gamma\tau(-e^{\gamma(\tau+C)})}{\left(1 - e^{\gamma(\tau+C)}\right)^2} = 0.$$

This equation holds if the denominator equals zero. I.e.,

$$1 - e^{\gamma(\tau+C)} - \gamma\tau(-e^{\gamma(\tau+C)}) = e^{\gamma\tau} \cdot e^{\gamma C}(1 - \gamma\tau) - 1 = 0.$$

Expanding the exponential function $e^{\gamma\tau}$ to the second degree we obtain $1 - e^{\gamma C} = \frac{1}{2}\gamma^2\tau^2$. Assuming that the time to save a checkpoint C is much shorter than the time between failures, i.e. $C \ll 1/\gamma$, which is a realistic assumption, a second order approximation to the term $e^{-\gamma C}$ gives $\tau^2 = \frac{2C}{\gamma} - C^2$. Ignoring the second order term gives the well known simple approximation [181].

$$\tau^* \approx \sqrt{\frac{2C}{\gamma}} \tag{9.5}$$

It should be noticed that this optimal time between checkpoints is not designed to minimise the expected completion time, but to minimise the overhead due to check-pointing and failures, certainly a related metric. Furthermore, (9.5) is an approxima-tion.

For systems in which failures may happen during normal operation or recovery, but no failures during checkpoint save operation, a very elegant derivation of the optimum checkpointing interval is given in [16]. It results in an exact solution for the optimum checkpointing interval,

$$\tau^* = \sqrt{C\left(2(\frac{1}{\gamma} - D) + C\right)} - C. \tag{9.6}$$

If the checkpointing cost C and the rollback time D are small compared with the time between failures $1/\gamma$ then (9.5) is a good approximation of the optimal check-pointing interval. The approximation overestimates the correct optimum checkpoint interval. In both the exact formulation as well as the approximation the indepen-dence of the task length should be pointed out.

In [93] the considered model can use a general distribution for the time to fail-ure. As a consequence the interval until the first checkpoint must be considered separately, since it is unlikely that this interval starts exactly with the beginning of an up time. Because of the memoryless property for exponentially distributed times between failures the probability of failure in the first interval in that case is just computed as in all other intervals. It is pointed out in [93] that the optimal checkpointing sequence that would minimise the expected completion time of a given program can only be computed if the time to failure is exponentially distributed. Then the optimisation corresponds to a resource allocation problem that can be solved using dynamic programming.

For non-exponentially distributed interfailure times the authors of [93] suggest an engineering approach, applicable only to equally long checkpointing intervals. They propose to compute the expected job completion time for different numbers of checkpoints and then select the number of checkpoints as optimal that yields the minimum expected completion time. In the case study a single checkpoint is able to improve expected completion time over no checkpointing, while adding more checkpoints adds too much overhead as compared to the failure characteristics. One would expect more benefit from checkpointing, which does, surprisingly, in the model not account for reprocessing after a failure.

In [30] checkpointing strategies are developed. A strategy consists of the number k of checkpoints to be placed and the amount of work $S_1, S_2, \ldots, S_k$ to be performed between checkpoints. The time to save the system state, the action of taking a checkpoint, is assumed to be constant and to take one unit of time. The simplest checkpointing strategy is the uniform strategy where equally spaced checkpoints are placed. Then only the number of checkpoints k must be chosen from which the interval length follows as

$$S_i = \frac{w}{k+1}$$

where w is the total amount of work to be performed. This very simple strategy has shown to be optimal for task restart in Chap. 5 where the purpose was to meet a deadline. One cannot expect the same property here although it will be shown for random checkpointing intervals that some metrics are optimised with equidistant checkpoints when not considering the last interval until job completion [30].

9.1.1.1 Models with Bounded Downtime

In this subsection the models from Chap. 2 with bounds on downtime are extended to include checkpointing. The considered application field are again critical tasks as investigated in [60] and [51] where either individual downtimes are bounded, the accumulated downtime may not exceed a bound or the number of accepted downtimes is bounded.

All model parameters are used as defined above and the system model is the simple two-state model depicted in Fig. 2.1 on p. 13. As in [51] we first extend the previous model by introducing checkpointing and neglecting the checkpointing overhead. Let us first consider the case of bounded individual downtimes.

We wish to compute the probability of completing the given work requirement w, $Pr\{w\}$. Assuming N down periods until job completion (and an equal number of up periods), recall that in preemptive repeat failure mode the probability of job completion (refer to (2.23)) is computed as

$$P(w|N = n) = 1 - (F_U(w))^n \qquad (9.7)$$

whereas in the preemptive resume failure mode (refer to (2.20)) the probability of job completion using n up times equals

$$\Pr\{w|N = n\} = 1 - F_U^{(n)}(w). \qquad (9.8)$$

Following [51] we will first add one checkpoint, which divides the task of length w into two sub-tasks of length $w/2$ each and then we will generalise to k checkpoints. Remember that we assume zero checkpointing cost. Given one checkpoint we derive a new expression for the probability of job completion $\Pr\{w|N = n\}$ using n up times.

Let n_1 be the number of up periods whose total duration falls into the interval $[0, w/2)$. These up periods are too short for checkpointing the performed work. Therefore, the work completed in the n_1 very short up periods is lost. This event occurs with probability $F_U(w/2)$. For illustration of the terminology see Fig. 9.1. Similarly, let n_2 be the number of longer up periods whose duration falls into the interval $[w/2, w)$, i.e. they are at least $w/2$ long and at most w. So, these intervals are always long enough to reach the checkpoint, in the best case the whole job can be completed in such an interval. In any case half of the necessary work is performed and with two such up times the job can be completed. This event occurs

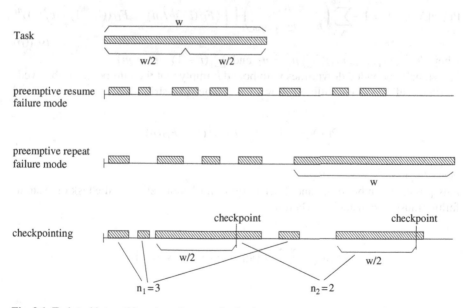

Fig. 9.1 Task and interval lengths using one checkpoint

with probability $F_U(w) - F_U(w/2)$. The job cannot be completed iff $n_2 < 2$ and all up times are less than w, that is, $n_1 + n_2 = n$. Given n up times the probability of task completion evaluates to

$$\Pr\{w|N = n\} = 1 - \sum_{(n_1,n_2)\in S} \binom{n}{n_1} (F_U(w/2))^{n_1} (F_U(w) - F_U(w/2))^{n_2} \quad (9.9)$$

where $S \equiv \{(n_1, n_2)|n_1 + n_2 = n, \text{ and } n_2 < 2\}$.

This expression is easily extended to the general case of k checkpoints. Then, a task is completed if its m subtasks of length w/m can be performed successfully, any subset of which can occur in the same up period. We assume that checkpoints are always placed between two subtasks.

Let n_i denote the number of up times whose durations fall in the interval $[(i - 1)w/m, i \cdot w/m)$ for $i = 1, \ldots, m$. The task cannot be completed iff

$$\sum_{i=1}^{m} n_i = n \quad \text{and} \quad \sum_{i=1}^{m} (i - 1)n_i < m.$$

Hence the probability of task completion using k checkpoints (corresponding to $m = k$ subtasks) and given n up periods is

$$\Pr\{w|N = n\} = 1 - \sum_{S} \binom{n}{n_1 \cdot n_2 \ldots n_m} \prod_{i=1}^{m} (F_U(i \cdot w/m) - F_U((i - 1) \cdot w/m))^{n_i},$$
$$(9.10)$$

where $S \equiv \{(n_1, \ldots, n_m)| \sum_i n_i = n, \text{ and } \sum_i (i - 1) \cdot n_i < m\}$.
Individually bounded downtimes with bound b imply that the number n of observed up times before system failure follows the geometric distribution:

$$\Pr\{N = n\} = F_D(b)^{n-1}(1 - F_D(b))$$

(see (2.9)).
Any possible number of up and down periods until completion of the task or system failure must be considered. Therefore

$$\Pr\{w\} = \sum_{n=1}^{\infty} \Pr\{w|N = n\} \Pr\{N = n\}.$$

When using the first M terms of the sum, the error is bounded by $(F_D(b))^M$

$$\Pr\{w\} = \sum_{n=1}^{M} \left(1 - \sum_{S} \binom{n}{n_1 \cdot n_2 \ldots n_m} \prod_{i=1}^{m} (F_U(i \cdot w/m) - F_U((i-1) \cdot w/m))^{n_i} \cdot \right.$$

$$\left. F_D(b)^{n-1}(1 - F_D(b)) \right) \quad (9.11)$$

For exponentially distributed up and down periods the probability of task completion is

$$\Pr\{w\} = \sum_{n=1}^{M} \left(1 - \sum_{S} \binom{n}{n_1 \cdot n_2 \ldots n_m} \prod_{i=1}^{m} \left(e^{-vb} \cdot \left(1 - e^{-vb}\right)^{n-1} \cdot \left(e^{-\gamma(i-1)w/m} - e^{-\gamma i w/m}\right)\right)\right)$$

$$(9.12)$$

When limiting the sum to only M down periods being considered, then the error can be bounded by $(F_D(b))^M$. This, however, is not a very tight bound, since for $vb = 5$ $F_D(b) \approx 1$ and very many down periods must be considered. In the analysis in this chapter vb ranges from 0 to 10 and the bound is not very useful.

Cost can be added to this model easily by replacing w/m with $w/m + c$. For the curves in Fig. 9.2 the added cost is 5% of the task length.

We wish to compare the probability of completing a task with checkpointing with the probability of completing a task in the preemptive repeat and the preemptive resume failure modes.

Recall from (2.20) the probability of task completion in preemptive resume failure mode

$$\Pr\{w\} = 1 - F_U^{(n)}(w)$$

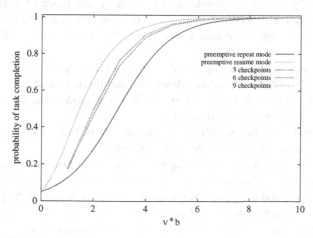

Fig. 9.2 Probability of task completion with and without checkpointing

which for an exponentially distributed system lifetime evaluates to

$$\Pr\{w\} = e^{-\gamma w e^{-vb}}. \tag{9.13}$$

In preemptive repeat failure mode (2.24) defines the probability of task completion which is

$$\Pr\{w\} = \frac{1 - F_U(w)}{1 - F_D(b)F_U(w)},$$

and for exponentially distributed up- and downtimes evaluates to (2.25)

$$\Pr\{w\} = \frac{1}{1 - (1 - e^{\gamma w})e^{-vb}}. \tag{9.14}$$

Figure 9.2 compares the preemptive resume, preemptive repeat failure mode and checkpointing with using the indicated number of checkpoints and associated cost of 5% of the job length. Only the curves with checkpointing are new, the curves for preemptive repeat and preemptive resume failure mode were shown already in Fig. 2.4 on p. 22.

In [51] the results for probability of task completion with checkpointing are given as numbers in a table and the authors point out that a different optimal number of checkpoints exist, depending on vb. When looking at the curves, this difference appears relatively so small, that it is of little significance.

The figure shows that preemptive repeat failure mode, which corresponds to one checkpoint at the beginning of the task, is a lower bound for checkpointing, while preemptive resume failure mode can be considered the perfect checkpointing strategy, a checkpoint is taken exactly when the system fails. So, the probability of task completion with checkpointing lies between those two extremes. It is remarkable, however, that the number of checkpoints does not make much of a difference at all. One would rather expect that the more checkpoints the closer the checkpointing curve is to the preemptive resume curve. Also, the insight obtained from [51] is limited.

The model can be extended by investigating different types of limitations of the downtime, i.e. bounded accumulated downtime and limited number of downtimes. This has not been done in the literature. Perhaps the model was not of sufficient practical value to attract further research.

Without bounds on the downtime the probability of task completion is not as interesting, because eventually, every task will complete.

In [63] different checkpointing strategies for equidistant checkpointing and Markovian checkpointing models are derived. It is shown that checkpointing adds extra work to the system and therefore always increases the job length, if the job length is assumed exponentially distributed. Checkpointing does, however, reduce the variance of job completion time, giving therefore more confidence in completion and establishing upper bounds on completion time. This property has been observed for restart as well, where it is independent of the distribution of task completion time (see Chap. 4 on p. 92).

9.1.1.2 Models for Forked Checkpointing

As described in Sect. 8 in forked checkpointing a child process is used to perform the checkpoint save operation. We assume checkpointing has a constant overhead, or cost, of C, the useful work performed between checkpoints is τ, the roll back cost equals R and as illustrated in Fig. 8.2 on p. 173 checkpoint latency is denoted L. The amount of useful work performed during checkpoint operation is $L - C$. If a failure happens after checkpoint C_1 and before checkpoint C_2 the system must perform R, the rollback cost, then it must repeat the work performed during the previous checkpointing operation, $L - C$, since the checkpoint always saves the state at the beginning of the latency period and, finally, an additional $\tau + C$ work must be done. This amounts in total to $R + (L - C) + (\tau + C) = R + \tau + C$ work that needs to be performed after a failure to complete the current interval.

Forked checkpointing does not directly optimise the expected task completion time, but a metric based on the expected task completion time, the *overhead ratio*. Let $E[T(w)]$ be the expected completion time of the whole job and $E[T(\tau)]$ the expected completion time of the portion τ, then the overhead ratio is defined as

$$r = \lim_{\tau \to \infty} \frac{E[T(\tau)] - \tau}{\tau}. \tag{9.15}$$

The expected completion time of the portion of work τ performed within one checkpoint interval is analysed in [156] using a three-state Markov chain as shown in Fig. 9.3.

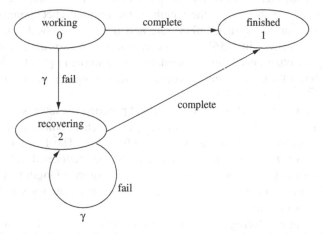

Fig. 9.3 Markov chain for evaluating the expected work performed per checkpoint interval

State 0 describes normal operation. When the model enters State 1 from state 0 the system has completed the interval without failure, a checkpoint has been established and amount τ of work has been performed. A failure at rate γ during the checkpoint interval is modelled as a transition to State 2. If only one failure happens, the transition to State 1 denotes completion of the interval. If, on the other hand, more failures occur then the model uses the self loop and remains in State 2.

Using Markov chain analysis [154] an expression for the expected completion time of an interval can be computed as the expected cost of a path from State 0 to State 1 in the three-state Markov chain. Each transition between states of the Markov chain has associated probability and cost. The cost corresponds to the expected time spent in one state before moving on to the next state. The expected completion time of an interval is computed as the expected cost of a path from State 0 to State 1. The expression obtained in [156] is

$$E[T(\tau)] = \frac{e^{\gamma(L-C+R)}(e^{\gamma(\tau+C)} - 1)}{\gamma} \tag{9.16}$$

Then the overhead ratio is

$$r = \frac{e^{\gamma(L-C+R)}(e^{\gamma(\tau+C)} - 1)}{\gamma \cdot \tau} - 1. \tag{9.17}$$

To optimise the overhead ratio with respect to the length of the checkpointing interval τ the derivative of the overhead ratio with respect to τ must be equal to zero:

$$\frac{\partial r}{\partial \tau} = \frac{\partial}{\partial \tau}\left(\frac{e^{\gamma(\tau+C)} - 1}{\tau}\right) = 0. \tag{9.18}$$

As shown in [156] there exists only one positive $\tau = \tau^*$ that satisfies this equation. Equation (9.18) shows that the length of the optimal checkpointing interval τ^* depends only on the system failure rate and the cost of checkpointing. Even though stated otherwise in [156] this observation is not surprising and has been made already for other checkpointing models. For example in [181] and [23] it has been shown that if the checkpointing cost C is small compared to the failure rate γ then $\tau^* \approx \sqrt{2C/\gamma}$. (see also Eq. (9.5)).

As can be seen in Fig. 8.1 for sequential checkpointing the latency is equal to the checkpointing cost and is therefore minimal. Forked checkpointing aims at time-sharing the checkpoint save operation with useful work. The fact that during the checkpoint save operation useful work can be performed reduces the checkpointing cost. In [156] it has been shown that the checkpoint interval length that minimises the overhead ratio is achieved together with the maximum latency which will also reduce the overhead ratio.

If a failure happens during checkpointing, recovery always rolls the system back to the most recent checkpoint. In forked checkpointing there is a chance that only the parent process has failed and the child process could save its checkpoint successfully. In this case the system can roll back to the actual checkpoint rather than to the previous one. This leads to faster recovery and the roll back model can be considered *optimistic*. Assuming that with the parent process the child process will always fail too gives rise to a *pessimistic* recovery model. As expected, in [70] the expected completion time of a task under optimistic recovery is shorter than when using pessimistic recovery. In forked checkpointing if the checkpoint interval τ is larger than the checkpoint cost C, which usually should be the case, then it pays off to wait for the checkpoint save operation to complete before rolling back the process.

9.1.2 Checkpointing Real-Time Tasks

The main characteristic of models for real-time tasks is the special treatment of failures. Real-time tasks are not only required to complete, but also to complete fault-free within a pre-defined time interval. Therefore, fault-detection and fault-treatments are handled in more detail than usually in the models in [137].

The basic model in [137] assumes a task length w with required execution time $T(w)$ and expected execution time $E[T(w)]$. The system fails at rate γ with failures occurring in a Poisson process. There are a total of K checkpoints inserted at processing time T_k, $1 \le k \le K$. The k-th checkpoint is inserted when the task has been successfully processed up to T_k. We define the work interval between checkpoints I_k, $0 \le k \le K$ as the computation time between the kth and the $(k+1)$th checkpoint, excluding the time t_c needed to save the checkpoint, i.e.

$$I_k = T_{k+1} - T_k - t_c.$$

The checkpointing interval includes saving a checkpoint as shown on the time line in Fig. 9.4 and is denoted as

$$\tau_k = I_k + t_c = T_{k+1} - T_k.$$

Surprisingly, in this model even after the task is completed a last checkpoint is taken. To avoid the unnecessary checkpoint, the authors in [30] point out that a checkpoint should be taken only if the remaining work is more than the checkpointing cost.

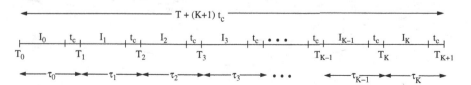

Fig. 9.4 Checkpointing intervals

An important assumption is that failures are detected immediately as they occur. More realistic would be a random delay in failure detection and a probability of not detecting the failure at all (which is equivalent to an infinite random detection delay). Modelling the failure detection in more detail requires either knowledge or assumptions on the characteristics of failure detection. As no data on the failure detection process is available additional assumptions would complicate the model while not providing more meaningful results.

The system can only keep one checkpoint in storage. If the rollback operation fails, because of a broken link to the checkpoint saving device or an internal failure of the checkpoint saving device the task has to be restarted from its beginning. Rollback succeeds with probability p and fails with probability $1 - p$ (in which case the task is restarted). If a failure happens before the first checkpoint the task always has to be restarted. The authors in [137] elaborate on the coverage of the different types of failures of the processor and the storage device. They introduce several parameters for coverage through an acceptance test and on-line failure detection which complicates the model and is ignored in the first model, which we present here.

Rollback of the system takes r time units while a restart of the task takes time s. It should be observed that repair of the system is not modelled explicitly, but rather implicitly within the rollback and restart time.

Figure 9.5 illustrates the two options. In state T_k time T_k is reached and interval τ_k, consisting of processing time I_k and checkpoint saving time t_c is just to begin. If within the interval τ_k a failure occurs then the system either rolls back to state T_{k-1} or it restarts again from T_0.

By going backwards through the intervals in Fig. 9.4 a recursive expression is derived in [137] for $\mathrm{E}\,[T(w)]$, which can be simplified into

$$
\mathrm{E}[T(w)] = hy_0u^K + vy \sum_{k=0}^{K-1} u^k,
\tag{9.19}
$$

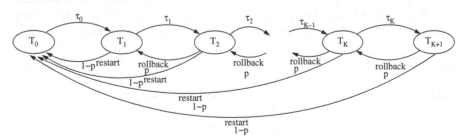

Fig. 9.5 Performing the amount of work $w = \sum_{i=0}^{n+1} I_i$ subject to rollback and restart

where

$$h = \frac{1}{\gamma} + s, \qquad\qquad v = \frac{1}{\gamma} + pr + (1-p)s$$
$$\tau^* = \frac{w-b}{n+1} + t_c, \qquad b = \frac{1}{\gamma} \ln\left(\frac{1+\gamma r}{1+\gamma s}\right)$$
$$y_0 = e^{\gamma b} e^{\gamma \tau^*} - 1 \qquad\qquad\qquad\qquad (9.20)$$
$$y = e^{\gamma \tau^*} - 1$$
$$u = (1-p)e^{\gamma \tau^*} + p.$$

$E[T(w)]$ must be minimised with respect to K and τ_k. A slight simplification could ease this step considerably. If the first interval would be treated the same as all other intervals then the optimum solution would have only equally spaced intervals and for a given task length the optimum number of checkpoint intervals can be determined. Once the optimum number of checkpoints is known the interval length for a task of given length can easily be determined. Similarly, for given interval length the number of checkpoints is known.

In the given model first the number of checkpoints is assumed fixed to minimise $E[T(w)]$ with respect to τ_i, then the above expressions are used to minimise $E[T(w)]$ with respect to K.

The following theorem relates the number of checkpoints K with the minimum expected completion time $E[T(w)]$. For the proof see [137].

Theorem 9.1 *For a given K the minimum $E[T(w)]$ is obtained when*

$$\tau_0 - \frac{1}{\gamma} \ln\left(\frac{1+\gamma r}{1+\gamma s}\right) = \tau_1 = \tau_2 = \ldots = \tau_K. \qquad (9.21)$$

This theorem shows that the optimal checkpoint intervals are equally long, except for the first one. This is not surprising, since in the first interval only restart is possible, while in all other intervals either rollback or restart can be performed.

To minimise $E[T(w)]$ with respect to K two cases must be distinguished. For the rollback probability holds either $p = 1$ or $p < 1$.

$$E[T(w)] = \begin{cases} h y_0 + nvy & \text{if } p = 1 \\ u^K \left(\frac{v}{1-p} + h y_0\right) - \frac{v}{1-p} & \text{if } 0 \le p \le 1 \end{cases} \qquad (9.22)$$

using the abbreviations in (9.20). The optimal K that minimises $E[T(w)]$ can be found by solving

$$\frac{d\, E[T(w)]}{d\, K} = 0 \qquad (9.23)$$

and finding the closest integer K. Unfortunately, no closed-form solution for (9.23) exists and the optimal K can only be found in an trial-and-error fashion by solving

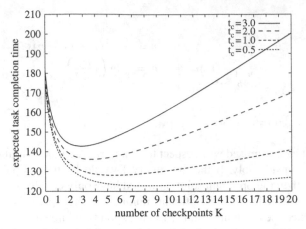

Fig. 9.6 Expected completion time versus number of checkpoints for $\gamma = 0.01, r = 0.2, s = 0.5,$ $p = 0.8, T = 100$

9.22 for successive K and finding its minimum. This procedure has been carried out in [137] and the results are illustrated in the following figures.

First the optimal number of checkpoints is evaluated with respect to the check-pointing cost. Figure 9.6 shows individual curves for different checkpointing cost. For the chosen parameters as indicated in the caption of Fig. 9.6 the precise optimum number of checkpoints is determined.

When generalising the observations made in Fig. 9.6, the rather obvious conclusion is that the lower the checkpointing cost, the lower the expected execution time while using more checkpoints. It is interesting to see that a task of length 100 with a failure rate of one per hundred time units and rather short rollback and restart time has a minimum expected execution time of 125% of its actual processing time. This is because only one checkpoint is being saved. In many cases, if the rollback fails, a considerable amount of work must be reprocessed. Adding more storage capacity could lead to a performance improvement for such systems.

Similar results are found in Fig. 9.7 when considering different failure rates.

With a failure rate of 0.001 on the average only one in 10 tasks experiences a failure. In this case checkpointing only adds to the expected task completion time and is not beneficial at all. The higher the failure rate, the higher becomes the expected task completion time and the more checkpoints should be used in order to recover from a failure with as little as possible lost work.

Figure 9.8 investigates the impact of the probability p of successful rollback. The optimum number of checkpoints is almost independent of the probability of successful rollback. Fewer checkpoints should be used as the probability increases that a complete restart of the task is necessary. This conforms to most people's intuition. Not surprisingly, the probability of successful rollback, or rather the probability of unsuccessful rollback has a significant impact on the expected task completion time.

This is because a restart of the task can be initiated at late stages in task execution and then comes with a large penalty.

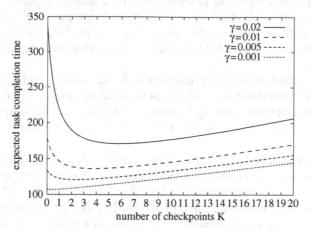

Fig. 9.7 Expected completion time versus number of checkpoints for $r = 0.2$, $s = 0.5$, $t_c = 2.0$, $p = 0.8$, $T = 100$

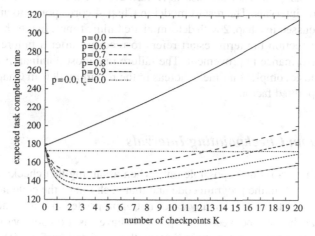

Fig. 9.8 Expected completion time versus number of checkpoints for $\gamma = 0.01$ $r = 0.2$, $s = 0.5$, $t_c = 2.0$, $T = 100$

The special case of $p = 0$ deserves some attention. If $p = 0$, i.e. rollback never takes place and the system restarts with each failure. Then, a task is restarted after an exponentially distributed processing time, instead of a fixed interval, as it was used for the restart model. The expected completion time in that case is a special case of (9.22), i.e.

$$E[T(w)] = \left(\frac{1}{\gamma} + s\right)\left(e^{\gamma(b+(K+1)\tau^*)} - 1\right)$$

where b and τ^* are as given in (9.20). If checkpointing is performed but rollback is always unsuccessful, such that a restart is necessary with each failure, then the checkpointing cost adds to the task execution time and the expected task execution time suffers from checkpointing. The expected task execution time with zero rollback probability is shown as the straight line with positive slope in Fig. 9.8. If, additionally, no cost is associated with checkpointing, then the expected completion time is constant.

As a rule of thumb from the figures can be deduced that one should rather take too many checkpoints than too few. All curves increase rapidly for fewer checkpoints than the optimum and more slowly for more checkpoints than the optimum number of checkpoints. The same engineering rule has been derived already for the restart and rejuvenation intervals.

If essentially no checkpoints are inserted and therefore the checkpointing cost is almost zero, then the number of checkpoints is irrelevant for the expected task completion time, which then only depends on the system failure rate. The respective curve is the straight horizontal line in Fig. 9.8. In this case the model only does restart. However, the model dynamics correspond to a model of completion time in unreliable systems as discussed in Chap. 2 rather than to the restart model in Chap. 4. Even though restart is performed, the restart does not happen at intervals chosen by the user. Instead, the task is restarted when the system fails, which is at random time intervals. The restart model in Chap. 4 corresponds to an unreliable system as discussed in Chap. 2 with deterministic failure times. In the checkpointing model of this section the term restart refers to recovery after a failure and not to restart for performance improvement. The failure rate most dominantly determines the expected task completion time, whereas in the restart models in Chap. 4 failures are not an appointed factor.

9.1.3 Random Checkpointing Intervals

Equidistant checkpointing is possible only when assuming that checkpoints can be placed at any point in the program code and that each run of the code takes equally long. Both assumptions are in many cases not appropriate, so it is a natural choice to assume that the time between checkpoints behaves as a random variable τ, following some probability distribution $F_\tau(t)$ with associated LST $F_\tau^\sim(t)$. The pieces of code of different length can be blocks or modules and therefore in [113, 90] this strategy is called *modular checkpointing*. It should not be confused with *random checkpointing* in [113] denoting a slightly more complex model. The random checkpointing model has the three system states *operational*, *in repair* and *checkpointing*. Since failures can happen during checkpointing the random variable denoting the length of the checkpointing interval is the minimum of the time until the next failure (if this happens during checkpointing) and the time needed for the actual checkpointing.

We will now discuss the model with the checkpointing intervals of random length and look at optimal checkpointing strategies as derived in [30].

In [113] two closely related measures are considered, the expected completion time of a program consisting of n modules and the expected processing time needed to finish a work requirement w. Both are generalisations of the above expected completion time when using equidistant checkpoints.

The expected completion time of a program that consists of n modules is

$$E[T(n)] = \left(\frac{1}{\gamma} + E[D]\right)\left((n-1)\cdot(F_{\widetilde{C}}(-\gamma)F_{\widetilde{\tau}}(-\gamma) - 1) + \left(F_{\widetilde{\tau}}(-\gamma) - 1\right)\right).$$
(9.24)

Equation (9.24) is obtained from (9.2) by replacing the equidistantly spaced time between checkpoints w/n with the random time τ and unconditioning on τ.

Obviously, for deterministic productive times between checkpoints of length $\tau = w/n$ we obtain $F_{\widetilde{\tau}}(-\gamma) = e^{\gamma w/n}$ and (9.24) reduces to (9.2). For exponentially distributed productive time between checkpoints with rate α $F_{\widetilde{\tau}}(-\gamma) = \frac{1}{\alpha-\gamma}$ (9.24) reduces to

$$E[T(n)] = \left(\frac{1}{\gamma} + E[D]\right)\left((n-1)\cdot\left(\frac{F_{\widetilde{C}}(-\gamma)}{\alpha-\gamma} - 1\right) + \left(\frac{1}{\alpha-\gamma} - 1\right)\right).$$
(9.25)

The above equations cannot be easily used for optimising the expected job completion time with respect to the number of checkpoints $K = n - 1$, since changing n also involves changing the distribution of the work requirement of each module $F_\tau(t)$.

We now consider the completion time of a given amount of work where a checkpoint is placed at the end of each module and the productive time between checkpoints is an iid random variable, rather than deterministic.

With generally distributed productive time between checkpoints the expected completion time of work requirement w in the presence of failures is [113]

$$E[T(w)] = \left(\frac{1}{\gamma} + E[D]\right)\left(F_{\widetilde{C}}(-\gamma)\int_{h=0}^{w} e^{\gamma h}\, dF_\tau(h) + e^{\gamma h}(1 - F_\tau(h)) - 1\right) +$$
$$\int_{h=0}^{w} E[T(w-h)]\, dF_\tau(h). \quad (9.26)$$

For the special case of deterministic checkpoint intervals (9.26) reduces to (9.2). For exponentially distributed productive time between checkpoints at rate α an explicit solution of (9.26) is given by

$$E[T(w)] = \left(\frac{1}{\gamma} + E[D]\right)\left(\frac{\gamma + \alpha(F_{\widetilde{C}}(-\gamma) - 1)}{(\alpha-\gamma)^2}\right) \cdot \left(\alpha(\alpha-\gamma)w + \gamma(e^{-(\alpha-\gamma)w} - 1)\right).$$
(9.27)

For $\alpha = 0$, i.e. no checkpointing, $E[T(w)]$ reduces to

$$E[T(w)] = \left(\frac{1}{\gamma} + E[D]\right)\left(e^{\gamma w} - 1\right)$$

given also in (2.2). For sufficiently large w it can be shown that $E[T(w)]$ is a convex function[1] and that checkpointing is beneficial. The optimal checkpointing rate then is higher than the failure rate of the system. For large w, i.e. for long running jobs, and checkpointing rate greater than the failure rate, i.e. $\alpha > \gamma$ we can approximate $E[T(w)]$ as

$$E[T(w)] \approx \left(\frac{1}{\gamma} + E[D] \right) \left(\frac{\gamma + \alpha(F_{\widetilde{C}}(-\gamma) - 1)}{\alpha - \gamma} \right) \alpha w.$$

This approximation is linear in w. An approximation to the optimal checkpointing rate $\hat{\alpha}$ is independent of w but dependent on the failure rate γ and given by

$$\hat{\alpha} \approx \gamma \left(1 + \sqrt{\frac{F_{\widetilde{C}}(-\gamma)}{(F_{\widetilde{C}}(-\gamma) - 1)}} \right).$$

When substituting $\hat{\alpha}$ in the approximation above one obtains for the expected time to complete a task of length w with checkpointing

$$E[T(w)] \approx \left(\frac{1}{\gamma} + E[D] \right) \gamma w \left(1 + 2(F_{\widetilde{C}}(-\gamma) - 1) + 2\sqrt{F_{\widetilde{C}}(-\gamma)(F_{\widetilde{C}}(-\gamma) - 1)} \right).$$

Following [113] this approximation provides good estimates for a wide range of parameters. It still requires knowledge of the distribution of the time needed for checkpointing and its LST.

A special case has been studied in [37]. It falls into the class of models for equally spaced checkpoint intervals and fixed number of checkpoints as discussed in the previous section. The model discussed in [37] assumes failures to occur in a Poisson process. They are detected as they occur and they do not happen during checkpointing. Repair takes a constant time D. A constant number of K checkpoints is placed in equally spaced intervals such that the total work requirement w is covered by checkpoint intervals of length τ, i.e. $w = K\tau$. The sequence of failure times $t_1, t_2, \ldots$, forms a renewal process. Consequently, also the sequence of checkpoint locations forms a renewal process.

The expected task completion time with checkpointing $E[T_C(w)]$ is shown in [37] to be

$$E[T_C(w)] = \frac{w}{\tau} \left(C + \left(D + C + \frac{1}{\gamma} \right) (e^{\gamma w} - 1) \right). \qquad (9.28)$$

Using (2.8) the solution with respect to w or γ of the inequality

$$E[T_C(w)] < E[T(w)]$$

---

[1] For a discussion of convexity in checkpointing see [30]

determines whether or not checkpointing is beneficial for the given amount of work in a system with the given failure rate. Both solutions for w and γ require a numerical evaluation (or approximation) of the exponential function.

The expected task completion time with checkpointing is determined as

$$\frac{d\mathrm{E}\left[T_C(w)\right]}{d\tau} = 0$$

which evaluates to

$$e^{\gamma \tau^*}(\gamma \tau^* - 1) = \frac{\gamma C}{1 + \gamma(D + C) - 1.}$$

The approximate solution is

$$\tau^* \approx \sqrt{\frac{2C}{\gamma\left(1 + \gamma(R + C)\right)}}. \tag{9.29}$$

Experimental [37] and analytical [90] analysis show that checkpointing is not beneficial for very short tasks, which complies with our intuition. The relative increase of the expected completion time in systems subject to failures $\mathrm{E}\left[T(w)\right]/w$ is exponential in the task length whereas in systems with checkpointing $\mathrm{E}\left[T_C(w)\right]$ it is linear in the task length. Without checkpointing a dominant factor is the failure rate of the system whereas with checkpointing the relative importance of the failure rate is compensated and the expected completion time increases less dramatically with the failure rate.

9.1.4 Algorithms for Optimum Checkpoint Selection

This subsection explores algorithms to determine good checkpointing strategies. Not much work exists that comes close enough to practical applicability as to formulate an algorithm. First, an algorithm for a checkpointing scheme with random checkpoint intervals is presented in which the system model and the program to be checkpointed are defined as in the previous subsection. The second algorithm is applicable to programs consisting of a given number of tasks where checkpoints can only be placed between tasks.

In [30] the optimal number of checkpoints to complete a given work requirement w is investigated. The task length is given, the time to perform a save operation (take a checkpoint) is constant and takes one time unit, the repair time of the system can be deterministic and then it takes $1/\nu$ time units, where ν is the repair rate as used above. The repair time can also be generalised to be a random variable D with probability distribution function $F_D(t)$. Time between failures is a random variable U with probability distribution $F_U(t)$. With each repair as well as through rejuvenation during the checkpoint save operation the system is transformed into a state that is as good as new. Consequently, the age variable of the system is reset with each repair and each rejuvenation during the checkpoint save operation. The system failure time distribution then is the same after repair and after checkpointing. A checkpointing strategy consists of choosing the number of checkpoints K and the amount of work performed between checkpoints $S_1, \ldots, S_K$.

The time between checkpoints is

$$s_k = S_k + 1, \qquad k = 1, \ldots, K.$$

Checkpointing is only performed when the remaining work is more than the time needed to save the system state. Therefore the following condition holds

$$S_1 + S_2 + \ldots, +S_K < w - 1.$$

Failures can happen during normal operation and during a save, the checkpointing operation. The system model is again as shown in Fig. 2.1 on p. 13. In the following we consider the repair time D to be a random variable distributed according to the distribution function $F_D(t)$.

The mean time to finish a job without checkpointing ($K = 0$) is

$$\mathrm{E}\,[T(w)] = \psi(w) = \frac{\left(\frac{1}{\nu} F_U(w) + \int_0^w (1 - F_U(x))\,dx\right)}{1 - F_U(w)} \qquad (9.30)$$

Considering $K \geq 1$ checkpoints gives

$$\mathrm{E}\,[T(w)] = \psi(s_1) + \psi(s_2) + \ldots \psi(s_{K+1}), \qquad (9.31)$$

where

$$s_k = S_k + 1, \qquad k = 1, \ldots, K$$
$$s_{K+1} = t + K - s_1 - s_2 - \ldots - s_K = t - S_1 - S_2 - \ldots - S_K.$$

If the work requirement is greater than one ($w > 1$), then all checkpointing strategies can be bounded from below by the relative actual portion of work performed between two checkpoints, i.e.

$$\mathrm{E}\,[T(w)] \geq m(w - 1), \qquad (9.32)$$

where

$$m \text{ is the greatest lower bound (glb) of } \frac{\psi(s)}{s - 1}, \qquad 1 < s \leq w.$$

Checkpointing of duration 1 is included in time s. Due to failure and rollback it can take longer than s to perform the work requirement and save operation s. The time needed, $\psi(s)$, will always be more than the pure work requirement $s - 1$, therefore $m(w)$ is always greater than one. The system performs better as the bound approaches one. For a proof see [30].

If $K, S_1, \ldots, S_K$ can be chosen such that

$$\frac{\psi(s_k)}{s_k - 1}, \qquad \text{for } k = 1, \ldots, K + 1$$

are all close to m, then $E[T(w)]$ is close to the lower bound $m(w - 1)$ and the strategy is close to optimal.

Consider for example an exponentially distributed time to failure $F_U(t)$ with mean $1/\gamma$, then (9.30) evaluates to

$$\frac{\psi(s)}{s - 1} = \left(\frac{1}{\nu} + \frac{1}{\gamma}\right)\frac{e^{\gamma s} - 1}{s - 1} \tag{9.33}$$

This equation has a unique minimum at $s = s^*$, which is at the root of

$$\gamma s + e^{-\gamma s} = 1 + \gamma. \tag{9.34}$$

This expression has in many cases no real-value solution and is, therefore, often not of immediate use.

If the task length is less than the optimal checkpointing interval, i.e. $w \leq s^*$, then (9.32) becomes $E[T(w)] \geq \psi(w)$ and (9.30) shows that the checkpointing strategy with $K = 0$, i.e. no checkpointing, is optimal.

If the task length is greater than the optimal checkpointing interval, i.e. $w > s^*$, then choosing a strategy that sets $s_k \approx s^*$ for all $k = 1, \ldots, K + 1$ has $\psi(s_k)$ close to $m(s^* - 1)$ and $E[T(w)]$ close to $m(w - 1)$. Then (9.32) shows that the strategy is close to optimal. Since the sum of all checkpointing intervals must evaluate to $w + K$, the task length plus the saves, the s_k cannot be all near s^* unless K is chosen such that approximately

$$\frac{w + K}{K + 1} = \frac{w - 1}{K + 1} + 1 = s^*. \tag{9.35}$$

This strategy leaves a longer job at the end $s_{K+1} = w - S_1 - S_2 - \ldots - S_K \approx s^*$, making all other intervals equally long. The optimal strategy differs from the equidistant strategy only in the length of the job left after the last checkpoint.

An optimal checkpoint interval length as in (9.35) does not exist for all K. As mentioned above the checkpointing intervals must add up to the task length and very often they will not do so when set to the optimal interval length s^*. Another prerequisite for the existence of an optimal interval length s^* is convexity of the expected time to finish a portion S_k of work $\psi(s)$. It turns out that the optimal checkpointing strategy depends on the failure rate of the system but is independent of the repair rate, which is a constant in the formulas.

For uniformly distributed time to failure with $0 \leq U \leq X$ and constant repair time of $1/\nu = 1/2$ the optimal checkpointing interval can be expressed with the simple formula

$$s^* = \sqrt{2(X+1)} - 1.$$

Table 9.1 lists the optimal checkpointing intervals $s^* - 1$ as a function of the failure rate γ. The proportion of work performed between checkpoints is determined by evaluating (9.34). This equation has two solutions for every value of γ, a positive and a negative one. We only use the positive solution, since a negative checkpointing interval makes no sense.

Using the table one immediately sees that if the number of checkpoints is fixed, very often the interval length cannot be optimal. Assume a task of length $s^* - 1 = 100$ must be processed and the system has failure rate of 0.001. If we set the number of checkpoints $K > 2$ then obviously the checkpointing intervals must be chosen much smaller than the optimal interval length. Even when using 2 checkpoints, the first would be placed at $t = 44$ and the second at $t = 88$. Then the last interval would only be 12 time units long and consequently much shorter than the optimal length.

To determine an optimal checkpointing strategy one should therefore first determine the optimal interval length for the failure time distribution of the considered system using (9.30) or for exponentially distributed time between failures (9.34). Consequently, (9.35) should be reformulated into

$$n = \left\lfloor \frac{w-1}{s^*-1} - 1 \right\rfloor. \tag{9.36}$$

Then for the above case with task length 100 and failure rate $\gamma = 0.001$ one would first compute the optimal checkpointing interval length of 44. One can then see that

Table 9.1 Optimal checkpointing interval to minimise the expected completion time

γ	$s^* - 1$
0.001	44.06
0.002	30.96
0.005	19.34
0.01	13.48
0.02	9.345
0.05	5.676
0.1	3.832
0.2	2.534
0.5	1.397
1.0	0.8414
2.0	0.4738
5.0	0.1995

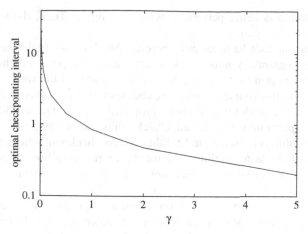

Fig. 9.9 System failure rate versus optimal interval length

the lowest expected completion time is achieved using only one checkpoint. This strategy chooses the last checkpoint interval longer than the previous one.

Figure 9.9 graphically depicts the data from Table 9.1. It shows the rapid growth of the optimal checkpointing interval length as the failure rate decreases and time between failures increases. Obviously, a system with no failures (i.e. failure rate $\gamma = 0$) needs no checkpoints, or infinite interval length.

As pointed out in [30] Eq. (9.34) does not always hold for an integer K. In fact, a necessary condition is that $\psi(s)$ is a convex function, which does not hold for all failure time distributions and needs to be proven for each considered distribution. Next to several examples of convex failure time distributions [30] considers a second model where no rejuvenation is performed during a checkpoint save operation. Hence the failure time distribution after repair is not the same as after a checkpoint save operation. After repair the system is as new while after a checkpoint save operation the system continues to degrade. This fact greatly complicates the treatment which we omit here. Instead, we refer the interested reader to [30].

The final part of this subsection is devoted to the algorithm presented in [152]. This is a prominent paper among the very few publications concerned with algorithms to determine checkpointing strategies. It is very interesting in the context of this book, since one of the major advantages of the simple restart model is that it serves as a basis to formulate algorithms to determine optimal restart intervals as to optimise a variety of metrics. As we will see, those algorithms have strong similarity.

The model in [152] assumes that the work to be processed is organised in a set of n modules, or a sequence of tasks, between which checkpoints can be placed, but not inside them. Therefore, the number of possible checkpoints is limited a priori by $n - 1$ and the potential location of checkpoint k is known to be between task $k - 1$ and task k. The model assumes constant failure rate. This can be the case e.g. if rejuvenation is being performed with each repair. Then, the system is 'as good as new' after a repair.

The problem can then be made more precise. An algorithm to determine an optimal checkpointing strategy must check which ones of the potential checkpoints to omit (if any) as to optimise the expected completion time of the whole sequence of tasks. The algorithm optimises cooperative checkpointing as analysed in [116, 135]. Strongly related is a model that has been proposed in [25], where a flowchart representation of a program is being used. Checkpoints can be placed at any arc in the graph. An algorithm is presented in [25] that selects checkpoints from the potential checkpoint locations as to minimise the maximum recovery time after a rollback. The most difficult part when using this approach in practice is the amount of knowledge required about the system recovery and rollback times.

If the considered system can be represented as a graph and the graph can be understood to represent a Markov chain, then, as shown in [102], the checkpointing policy that corresponds to a control limit policy is an optimal checkpointing strategy that can be found in a straight forward manner. A control limit policy R_k places checkpoints only in states with index greater or equal k. If there are N states and $k = 0, \ldots, N$ then R_0 can place a checkpoint in any state while R_N cannot place a checkpoint at all.

For two reasons the arbitrary placement of checkpoints in a job of length w, as illustrated above and in [30], is much more difficult to determine. First, the number of checkpoints in an arbitrary task is not necessarily limited and second, their potential location is not predefined.

The computation is formalised as follows. The sequence of n modules or tasks constitute together the considered program. The tasks are numbered as $i = 1, \ldots, n$. Task i has processing requirement t_i in absence of failures. The boundary between task $i - 1$ and task i is the i−th candidate checkpoint location with checkpointing cost c_i. for saving the system state before computation of task i starts. When a failure is detected the system rolls back to the most recent checkpoint at roll back cost r_i, if the most recent checkpoint is at location i. Figure 9.10 illustrates the notation. There is no checkpoint placed between Task 5 and Task 6, neither between Task 6 and Task 7. Therefore, the failure in Task 7 calls for a roll back to the beginning of Task 5 at roll back cost r_5. The initial situation is assumed to constitute a checkpoint before the first task, which is not explicitly shown in the figure. Assumptions that have later been relaxed in [152] are failure free checkpoint setup and failure free rollback operation.

The optimisation problem is to choose a subset of the $n - 1$ checkpoint candidates as to minimise the expected completion time of the whole task sequence.

Before presenting the algorithm that solves this optimisation problem some further notation is needed as illustrated in Fig. 9.10. A sequence consisting of the tasks $i, i + 1, \ldots, j - 1, j$ is denoted $[i, j]$. The sequence $[1, n]$ refers to the entire computation. Let $T_{i,j}^K$ define the expected completion time of the task sequence $[i, j]$ with at most K checkpoints. Clearly, $T_{1,m}^0$ refers to the expected completion time of the whole program without any checkpoints.

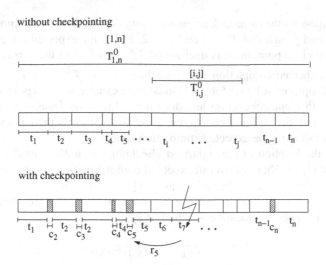

Fig. 9.10 Notation used for optimal checkpoint selection

The *K-optimal solutions* for the task sequence $[i, j]$ are those checkpoint selections in $[i, j]$ that achieve the minimal expected completion time $T_{i,j}^K$ with at most K checkpoints. The K-optimal solutions of $[i, j]$ with $k \le K$ checkpoints are denoted $\mathbf{L}_{i,j}^K$. If $\mathbf{L}_{i,j}^K$ contains no checkpoints it is written as $\mathbf{L}_{i,j}^K = <>$ while if $\mathbf{L}_{i,j}^K$ contains k checkpoints, the K-optimal solution is denoted as the sequence of checkpoint locations $\mathbf{L}_{i,j}^K = < u_1, u_2, \ldots, u_k >$, where $i < u_1 < u_2 < \ldots < u_k \le j$. The rightmost checkpoint location of $\mathbf{L}_{i,j}^K = < u_1, u_2, \ldots, u_k >$ is u_k, while the rightmost checkpoint location of $\mathbf{L}_{i,j}^K = <>$ is i, since there is no checkpoint besides the initial state.

Note that there can be several different checkpoint selections with the same number of checkpoints where the checkpoints are placed differently. A K-optimal solution must satisfy the additional condition that the rightmost checkpoint location in $\mathbf{L}_{i,j}^K$ is greater than or equal to the rightmost checkpoint location of any other K-optimal solution or $\mathbf{L}_{i,j}^K = <>$. With this definition, still more than one K-optimal solution can exist. For a real-valued expected completion time the event that several different checkpoint selections with the same number of checkpoints achieve the same expected completion time is very unlikely. Practical issues such as how to decide when two computations of the expected completion time with different checkpoint selections are identical is an open problem.

The checkpoint selection algorithm recursively chooses the rightmost checkpoint location to optimise the expected completion time using K checkpoint selections and leaves the problem with $K - 1$ checkpoint locations, for which again the rightmost is chosen and so on until the last checkpoint location is determined. In more detail the algorithm operates as follows. Let us consider the expected completion time $T_{1,j}^K$ of the task sequence $[1, j]$ using at most K checkpoints and the expected completion time for the same task sequence using at most $K - 1$ checkpoints $T_{1,j}^{K-1}$ for some K and j such that $K \geq 1$ and $j \geq 2$. Either the expected job completion time with one checkpoint more is unchanged $T_{1,j}^{K} = T_{1,j}^{K-1}$ or the extra checkpoint reduces the expected completion time and then $T_{1,j}^{K} < T_{1,j}^{K-1}$. If $T_{1,j}^{K} < T_{1,j}^{K-1}$, then every K-optimal solution for $[1, j]$ must have exactly K checkpoints, since we just assumed that one checkpoint less does not achieve the same expected overall completion time. So no other checkpoint selection with less than K checkpoints exists which has the same expected completion time.

Let h be the location of the rightmost checkpoint of a K-optimal solution for the sequence $[1, j]$. Then the overall expected completion time can be split into the expected completion time of the task sequence $[1, h - 1]$ using $K - 1$ checkpoints and the expected completion time of the last segment $[h, j]$ plus the checkpointing cost for the checkpoint at location h, which is c_h.

$$T_{1,j}^{K} = T_{1,h-1}^{K-1} + T_{h,j}^{0} + c_h.$$

In other words, up to $K - 1$ checkpoints have been established optimally in $[1, h - 1]$ and the last checkpoint is located at h. No more checkpoints are in the interval $[h, j]$. Let $\mathbf{L}_{1,h-1}^{K-1}$ be a $(K - 1)$-optimal solution for $[1, h - 1]$ then adding a checkpoint at location h

$$\mathbf{L}_{1,j}^{K} = \mathbf{L}_{1,h-1}^{K-1} || < h >$$

must be a K-optimal solution. The operator $||$ denotes concatenation of sequences, i.e. $< u_1, u_2 > || < u_3 > = < u_1, u_2, u_3 >$.

We can generalise this observation to compute $T_{1,j}^{K}$ and $\mathbf{L}_{1,j}^{K}$ as follows. Let

$$T = \min_{1 < i \leq j} (T_{1,i-1}^{K-1} + T_{i,j}^{0} + c_i)$$

and let h be the largest index i such that $T = T_{1,h-1}^{K-1} + T_{h,j}^{0} + c_h$. The expected completion of the sequence $[1, j]$ with K checkpoints must be the minimum of the sequences with K and $K - 1$ checkpoints $T_{1,j}^{K} = \min(T, T_{1,j}^{K-1})$. The K-optimal solutions can be defined as

$$\mathbf{L}_{1,j}^{K} = \begin{cases} \mathbf{L}_{1,j}^{K-1} & \text{if } T_{1,j}^{K} = T_{1,j}^{K-1} \\ \mathbf{L}_{1,h-1}^{K-1} || < h > & \text{else.} \end{cases}$$

The previous paragraph shows that if for all i and j such that $1 \leq i \leq j$ the expected completion time without checkpoints $T_{i,j}^{0}$ and the expected completion time with $K - 1$ checkpoints $T_{i,j}^{K-1}$ and the $(K - 1)$-optimal solutions $\mathbf{L}_{i,j}^{K-1}$ have been computed, then also $T_{i,j}^{K}$ and $\mathbf{L}_{i,j}^{K}$ can be determined. This is a recursive approach starting from the rightmost potential checkpoint location which lends itself for a dynamic programming algorithm. This algorithm is formalised in

Algorithm 9 (Computation of the Optimal Checkpoint Selection)

```
For i = 1 to n {
    For j = 1 to n {
        compute T⁰ᵢ,ⱼ;
    }
}

For K = 1 to n − 1 {
    Tᴷ₁,₁ = T⁰₁,₁;
    Lᴷ₁,₁ = < >;
}

For K = 1 to n − 1 {
    For j = n to 2 {
        T = min₁<ᵢ≤ⱼ (Tᴷ⁻¹₁,ᵢ₋₁ + T⁰ᵢ,ⱼ + cᵢ);
        Let h be the largest minimising index i;
        If T < Tᴷ⁻¹ᵢ,ⱼ then {
            Tᴷ₁,ⱼ = T;
            Lᴷ₁,ⱼ = Lᴷ⁻¹₁,ₕ₋₁ || < h >;
        }
        else {
            Tᴷ₁,ⱼ = Tᴷ⁻¹₁,ⱼ;
            Lᴷ₁,ⱼ = Lᴷ⁻¹₁,ⱼ;
        }
    }
}
```

It is worth mentioning that the failure model does not explicitly appear in the system. It is implicitly present, since the expected completion times of all partial sequences of tasks $T^0_{i,j}$, of course, depend on the failure model as well as on the completion time distribution. The first step in the algorithm is its weakest point. The computation of all $T^0_{i,j}$ has already computational complexity of order $O(m^2)$ and because of the minimum operator Algorithm 9 in total is of order $O(m^3)$.

The algorithm can be slightly modified in that the rollback costs are related to the checkpoint set up costs, such that the rollback cost is a linear function of the set up cost as $r_i = \alpha c_i + \beta$ for some constants $\alpha \geq 0$ and $\beta \geq 0$. Then if $c_i > c_j$ it follows that $r_i \geq r_j$ and the loops in the algorithm can be reduced and the complexity of the algorithm is only $O(m^2)$, see [152].

Two failure models are considered in [152]. A discrete Bernoulli model and a continuous model where failures constitute a renewal process.

The discrete failure model is outlined first. Although failures can happen at any time in task i, they are detected only at the end of task i. Each task has a probability p_i of completing successfully in time t_i. Note that p_i does not denote the failure probability, as usual, but the probability that no failure happens. In [152] the moment generating function is derived and used to determine a recurrence relation for the expected task sequence completion times without checkpoints as

$$T_{i,i}^0 = \frac{t_i}{p_i} + \left(\frac{1}{p_i} - 1\right) r_i \tag{9.37}$$

$$T_{i,j}^0 = \frac{1}{p_j} + \left(T_{i,j-1}^0 + t_j\right) + \left(\frac{1}{p_j} - 1\right) r_i \quad \forall j, j > i. \tag{9.38}$$

Using these recurrence relations all $T_{i,j}^0$ with $1 \le i \le j \le m$ can be computed in $O(m^2)$ time.

The continuous failure model uses the random variable X for the time between failures, which is independent and identically distributed. In this failure model, failures are detected immediately as they happen.

The time between failures, X, has a known probability distribution function, $F(x) = \Pr\{X \le x\}$. Then the expected completion time of a sequence of tasks without checkpoints can be expressed as

$$T_{i,j}^0 = t_{i,j} + \frac{r_i F(t_{i,j}) + \int_0^{t_{i,j}} t \, dF(t)}{1 - F(t_{i,j})}. \tag{9.39}$$

For the special case of exponentially distributed epochs between failures this evaluates to

$$T_{i,j}^0 = \frac{(e^{\gamma t_{i,j}} - 1)(\gamma r_i + 1)}{\gamma}.$$

Algorithm 9 is to some extent similar to the backward algorithm (Algorithm 2 on p. 62 on p. 62) for optimising the first moment of completion time under restart. Both algorithms are dynamic programming algorithms, both determine recursively backwards starting from the last interval the optimal intervals. The backward algorithm uses a simpler expression for the expected completion time of the segments and is therefore able to determine the optimal length of those segments. Algorithm 9 also determines the checkpoint interval length, by potentially omitting checkpoints, but it has much less degree of freedom. However, the model the backward algorithm operates on is also simpler and more restricted as it has only one checkpoint at the origin, the leftmost checkpoint in the checkpointing model.

An on-line algorithm for checkpoint selection is derived in [183]. This algorithm optimises the overhead ratio by minimising the checkpointing cost. It is assumed that checkpointing cost is dependent on the system state (e.g. through the amount of memory allocated) and if the system state can be monitored, checkpoint locations can be chosen such that the checkpointing cost is low. This minimises the checkpointing overhead. The system state is captured in a simple Markovian model. The algorithm is evaluated in a simulation study. The results are compared with the simulation of the optimal checkpoint selection algorithm from [152]. This alleviates a deficiency in [152], the lack of experimental results. The checkpointing overhead is minimal in the optimal checkpoint selection algorithm [152], it is close to optimal when using the online algorithm [183], which is still better than using equispaced checkpointing intervals.

Algorithms for checkpointing and rollback in distributed systems are proposed in [81]. Much work on coordinated checkpointing has been published over the last 30 years. Checkpointing in distributed systems is to a large extent concerned with synchronisation problems in distributed systems.

Two main conclusions should be drawn from the analysis of program level checkpointing. First, constant checkpoint intervals minimise the expected completion time of a task. It is interesting to note that the expected completion time of a task under restart is maximised using constant restart intervals. For higher moments of completion time under restart constant interval length is no longer optimal and the same is to be expected for checkpoint intervals, albeit no work on higher moments of completion time under checkpointing is known.

Second, the optimal checkpoint interval is independent of the task length. This does not hold for the optimal restart interval, which does depend on the moments of completion time. Also, in system level checkpointing the optimal checkpointing interval does depend on the load on the system, as we will see in the next section.

9.2 Checkpointing at System Level

While in program level checkpointing as well as in restart the focus is on long-running tasks that should complete as fast as possible, in system level checkpointing tasks are typically short and the system is sometimes seen as a queueing system where the states of the server are described using a stochastic process. The total processing time of a task is short with respect to the time between failures of the system. Consequently, task completion time is not a critical measure and system availability, or unavailability [63] is investigated instead. The optimal checkpointing intervals are not the same for minimum expected task completion time and for maximum expected system availability [53].

An exact formula for the optimum checkpoint interval in [52] shows that the optimum checkpoint interval in transaction-based systems depends on the system load. The approximation of the optimal checkpointing interval $T_{opt} \approx \sqrt{2C/\gamma}$ in [181, 23], where C is the checkpointing cost and γ the system failure rate, is still applicable to transaction-based systems if the system is saturated.

New to system level checkpointing, as compared to program level checkpointing, is the fact that establishing a checkpoint in transaction-based systems does not only require saving the system state, but also recording an audit trail of transactions that were processed. In case of a failure the system is rolled back to a functional state and the saved transactions are reexecuted. A typical example of a transaction-based system is a database. In database systems two types of transactions are distinguished. First, those transactions that merely query the data base and second, the transactions that create or change entries of the database.

9.2.1 Analytic Models for Checkpointing Transaction-Based Systems

In this subsection three models taken from [24] are presented. The models A, B, and C (as they are also referenced in later literature) have increasing complexity. All three models make the following four assumptions:

1. faults are detected at random times in a (possibly non-homogeneous) Poisson process. This is equivalent to the very common assumption of occurrence of failures in a Poisson process and immediate fault detection.
2. the time required to reprocess the audit trail is directly proportional to the number of transactions recorded in the audit trail. This also is a common assumption
3. Transactions that arrive while the system is checkpointing or recovering from a failure are being stored and processed later. The time required to process stored transactions is assumed small compared to the MTBF. Many papers assume that requests during checkpointing and recovery are rejected.
4. System availability under optimal checkpointing is assumed to be high. This issue is not usually addressed but the discussion of checkpointing strategies with respect to system availability constitutes a main contribution of [24].

Model A uses two additional assumptions:

1. all request rates are constant and
2. no failures happen during recovery.

The second assumption is widely used and reasonable if checkpointing and recovery time are small compared to the MTBF.

Model B removes the second assumption of Model A and allows for failures during checkpoint establishment and rollback recovery. Model C relaxes the first restriction in Model A using a seasonal request rate that varies over time. Cycles will typically be days or weeks. We will not discuss Model C here because it leads to a very complex description that requires numerical approximation which does not allow us to compute results which can be compared with the results from other models. The model is therefore of limited practical value and gives only little insight. In [92] a cyclic failure rate is used as well.

For both, Model A and Model B the following definitions are required. Let C be the checkpointing cost (which can be a time penalty induced by checkpointing) and R the time needed for rollback. The system recovery is divided into the rollback of the system and the reprocessing of the audit-trail. Let $r(t)$ be the expected system recovery time given a failure happened at time t after the most recent checkpoint, which is a monotone, non-decreasing function. Let τ be the time between checkpoints and let $R(\tau)$ be the total expected time spent in recovery between two checkpoints, given that the time between two checkpoints is τ. $R(\tau)$ is monotone, increasing with τ. The total overhead $O(\tau)$ consists of checkpointing and recovery within the checkpoint interval τ and

$$O(\tau) = C + R(\tau).$$

Define the expected relative overhead $o(\tau)$ as

$$o(\tau) = \frac{O(\tau)}{\tau}.$$

In [24] convexity of $o(\tau)$ is shown. Availability of the system can be expressed in terms of the checkpointing interval τ as

$$A(\tau) = 1 - o(\tau). \tag{9.40}$$

The failure rate of the system is γ and t counts the time since the most recent checkpoint. A checkpoint is initiated at time $t = 0$ and the next checkpoint is taken at $t = \tau$. The time needed to reprocess the audit trail in case of a failure at time t is directly proportional to t. The constant of proportionality is ρ, also called the *compression factor*. The transaction arrival rate is λ the processing rate of transactions in the audit trail is μ. Until time t a total expected number of λt transactions arrive and the required time to process those transactions is $\lambda t/\mu$. Therefore the compression factor ρ equals the utilisation of the queue

$$\rho = \frac{\lambda}{\mu}.$$

Considering only the transactions that modify the database, with respective arrival rate λ' and processing rate μ' the compression factor is $\rho = \lambda'/\mu'$. In stable systems the utilisation is less than one and therefore $\lambda < \mu$. Furthermore, the number of transactions that modify the data base is usually small and therefore μ' will be less than μ, similarly λ' will be less than λ and the compression factor will not change much. Values of $\rho = 1/10$ should be expected.

For Model A the expected recovery time $r(t)$, given a failure is detected t time units after the most recent checkpoint further explains the total expected time between checkpoints $R(\tau)$. The function $r(t)$ consists of the rollback time and the time required for reprocessing of the audit trail. Therefore,

$$r(t) = R + \rho t. \tag{9.41}$$

The expected cost of recovery in the interval $[t, \tau]$, i.e. from time t until the next checkpoint at time τ is denoted $C(t)$. Note that $C(\tau) = 0$ and $R(\tau) = C(0)$ is the total expected cost of recovery between two consecutive checkpoints.

In real systems the load on the system will not be constant. Even if the arrival rate of requests is constant due to checkpointing and rollback recovery after the period of checkpointing and after the period of rollback a large number of requests will have accumulated, shortly increasing the otherwise constant request rate.

For simplification, in Models A and B in [24] the request rate is assumed constant always and checkpointing and recovery are assumed to happen instantaneously. Checkpointing is associated with cost C and rollback has cost $r(t)$ if a failure happens t time after the most recent checkpoint. This simplification is reasonable only if system availability is high.

For Model A in [24] the following properties are proven. We only list and interpret the theorems here. For the proofs see [24].

Lemma 9.2 *The variable cost function $C(t)$ satisfies*

$$\frac{d\,C(t)}{dt} = -\gamma \cdot r(t), \qquad \text{for } 0 \le t \le \tau. \tag{9.42}$$

Intuitively, the increment of the expected cost function between two checkpoints depends on the cost of rollback and recovery and the frequency of its occurrence.

Theorem 9.3 *The optimal value of the intercheckpoint time τ^* which minimises the relative overhead $o(t)$ and maximises availability is*

$$\tau^* = \sqrt{\frac{2C}{\gamma\rho}} \tag{9.43}$$

with corresponding relative overhead

$$o_\tau^* = \gamma R + \sqrt{2C\gamma\rho}. \tag{9.44}$$

As the system becomes saturated and the utilisation $\rho \to 1$ the optimal checkpointing interval reduces to the widely known approximation

$$\tau^* = \sqrt{\frac{2C}{\gamma}} \tag{9.45}$$

It should be noticed that in this model task completion time is not being optimised. It is not even considered, because tasks are assumed to be short in relation with MTBF. Nonetheless, (9.45) is identical with (9.5) on p. 181, which has been derived as an approximation of the optimal checkpoint interval for equispaced checkpointing of long-running batch programs. Later, in [52], it has been shown, that the formula is exact only for saturated systems and otherwise is an approximation. The quality of the approximation is shown in the next section in Fig. 9.18 on p. 228. Therefore, (9.45) is widely used as a general rule to determine optimal checkpoint intervals, although it is in most cases an approximation.

Using $r(t) = R + \rho t$ and the boundary condition $C(\tau) = 0$ Eq. (9.42) can be solved and we obtain for the total expected time spent in reprocessing during a checkpoint interval of length τ and the relative expected time spent in overall recovery

$$R(\tau) = C(0) = \gamma \cdot R \cdot \tau + \frac{\gamma \cdot \rho \cdot \tau^2}{2} \qquad (9.46)$$

and

$$o(\tau) = \gamma \cdot R + \frac{\gamma \cdot \rho \cdot \tau}{2} + \frac{C}{\tau} \qquad (9.47)$$

Corollary 9.4 *At optimality, the maximum recovery time is*

$$r(\tau^*) = \frac{o(\tau^*)}{\gamma}. \qquad (9.48)$$

In other words, the maximum time spent in rollback and recovery equals the relative overhead due to rollback and recovery times the MTBF. If the system spends 10% of the time between checkpoints in rollback and recovery and the MTBF equals 100 then the maximum time for recovery is 10 time units.

Consequently, the expected number of failures during one individual recovery is $o(\tau^*) = 0.1$ in the example above. In [24] a relative recovery time of up to 10% (or availability of 90%) leads to a negligible number of failures during recovery which can be knowingly ignored and the simplification of no failures during recovery therefore can be tolerated.

Elasticity is a measure which indicates the sensitivity of a function towards changes in a variable. It consists in the ratio of the percentage change in the variable and the percentage change of the function using that variable. For a function $f(x)$ elasticity with respect to x is defined as

$$E_{f,x} = \left| \frac{\partial \ln f(x)}{\partial \ln x} \right| = \left| \frac{\partial f(x)}{\partial x} \frac{x}{f(x)} \right|.$$

For checkpointing the sensitivity of the checkpointing cost as well as the checkpoint interval with respect to the failure rate of the system is of importance. The failure rate of a system is extremely difficult to estimate precisely. Failures happen only rarely and therefore the sample size is always too small to obtain an estimate with high confidence.

Corollary 9.5 *The following bounds on elasticities exist*

$$\frac{\partial o(\gamma, t)}{\partial \gamma} \frac{\gamma}{o(\gamma, t)} \le 1$$
$$\frac{\partial o(R, t)}{\partial R} \frac{R}{o(R, t)} \le 1$$
$$\frac{\partial o(\rho, t)}{\partial \rho} \frac{\rho}{o(\rho, t)} \le \frac{1}{2}$$
$$\frac{\partial o(C, t)}{\partial C} \frac{C}{o(C, t)} \le \frac{1}{2}$$

(9.49)

The proof is by direct differentiation as in [24].

This corollary expresses the fact that a change of 1% in the checkpointing cost C implies a change of at most 0.5% in the expected relative overhead. In most cases the change of the expected relative overhead will be even less. This is important because a not too large error in the estimate of the system parameters does not render the model useless.

Figure 9.11 shows the relative overhead in Model A. The used parameter values are the cost of checkpointing $C = 0.5$, the rollback cost $R = 0.5$, system failure rate $\gamma = 1/18$ and compression factor $\rho = 1/8$. The total overhead consists of a constant part per interval, which is the time needed to take a checkpoint, and the variable part, which is the cost to reprocess the audit trail. The latter part is variable because the number of transactions to reprocess depends on the length of the checkpointing interval. The cost of taking a checkpoint is constant, but when considered relative to the interval length it decreases as the interval length increases.

The variable checkpointing cost in (9.46), on the other hand, increases proportional to the square of the checkpoint interval, therefore the relative rollback and recovery cost increases linearly with the checkpoint interval. The sum of both, the

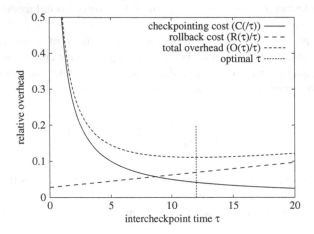

Fig. 9.11 Minimised checkpointing costs

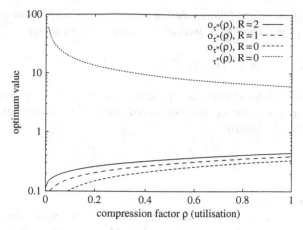

Fig. 9.12 Optimal checkpoint interval and relative minimal overhead

relative fixed and the relative variable checkpointing costs has its minimum where the optimal checkpointing interval τ is marked in Fig. 9.11. It coincides with the value obtained by evaluating (9.43). This figure very much resembles the figures for the rejuvenation interval in Chap. 6.2. Also, the issue of optimising a trade-off between the cost of a failure and the cost of the fault-tolerance mechanism is similar.

Figure 9.12 uses the same parameters as above (rollback cost $R = 0.5$, system failure rate $\gamma = 1/18$ and compression factor $\rho = 1/8$). The cost of checkpointing is $C = 1.0$. The figure displays the sensitivity of the relative overhead, as given in (9.44), and the optimal checkpoint interval, (9.43), to the compression factor ρ. As the compression factor increases either more transactions require reprocessing, or their processing time grows longer.

The optimal checkpoint interval decreases as the compression factor increases. A shorter checkpoint interval of course leads to lower reprocessing cost. The constant part of the rollback and recovery cost, the restoration R of the most recent checkpoint does have an influence on the optimum relative checkpointing cost. The optimum relative checkpoint overhead increases with the compression factor. Figure 9.12 uses a logarithmic scale, which flattens the pronounced increase.

Model B takes possible failures during checkpointing and recovery into consideration. Model B should be used if the system availability is below 90% as then a failure during recovery is rather likely. The difference between Model A and Model B concerns the definition of the recovery time $r(t)$, given a failure occurs t time after the most recent checkpoint (see (9.41)).

If failures can happen during recovery a recovery is completed only after $R + \rho t$ time units of failure-free reprocessing have taken place. If a failure happens during recovery the reprocessing is reinitiated. Several attempts of recovery might be necessary before reprocessing completes successfully. This requires a new definition of the optimal checkpoint interval, as given in the following theorem.

Theorem 9.6 *If there are failures during recovery, but no failures during check-pointing, the optimal checkpoint interval τ^* satisfies the equation*

$$e^{\gamma\rho\tau^*}(1 - \gamma\rho\tau^*) = 1 - \gamma\rho C \cdot e^{-\gamma R}. \tag{9.50}$$

If there can be failures during checkpointing and recovery and both the check-pointing and recovery process are reinitiated after each failure then the optimal checkpoint interval τ^* satisfies

$$e^{\gamma\rho\tau^*}(1 - \gamma\rho\tau^*) = 1 - \left(e^{\gamma C}01\right) \cdot \rho \cdot e^{\gamma R}. \tag{9.51}$$

For the proof see [24].

As the failure rate γ tends to zero, the optimal checkpoint interval of Model B tends to the optimal checkpoint interval of Model A. The optimal checkpoint interval in Model B is always shorter than that for Model A.

Obviously, all overhead is higher in Model B than in Model A. While for Model A a closed-form expression for the optimal checkpoint interval exists, this is not the case for Model B. The optimal checkpoint interval in Model B must be determined numerically.

9.2.2 Checkpointing Policies for Transaction-Based Systems

A very general and powerful model is used in [146]. The model allows for general distributions of the time between failures, instead of the commonly used exponential distribution. The checkpoint interval can depend on the reprocessing time and on the failure distribution. Furthermore, failures may occur during checkpointing and recovery.

The considered metric is the system availability. The usual equidistant check-pointing strategy depends on the mean of the failure time distribution. In [146] a failure-dependent but reprocessing-independent cost strategy is introduced, the *equicost* strategy. The equicost strategy uses the trade-off between the cost of check-pointing and the cost of recovery to determine an optimal checkpoint interval. For exponentially distributed times between failures both strategies achieve the same optimal checkpoint interval and the same optimal availability, while for other distributions of the time between failures this is not the case.

The model is defined by the failure process and the checkpointing process. We first describe the failure process.

Failures that happen during normal processing are denoted $F_k, k = 1, 2, \ldots$ while failures during recovery or reprocessing are denoted $f_k, k = 1, 2, \ldots$. The time of occurrence of a failure is $t(F_k)$ and $t(f_k)$, respectively. Then the time between failures $G_k = t(f_{k+1}) - t(f_k)$, where f_k is an arbitrary failure during either normal operation or recovery and reprocessing, constitutes a stationary stochastic process that is independent of the system state. Let the times between failures

$G_k, k = 1, 2, \ldots$ be independent and identically distributed with mean time between failures $1/\gamma$. The time between two consecutive failures during normal operation $L_i = t(F_{i+1}) - t(F_i)$ is the length L_i of the i-th cycle. The cycle length consists of two portions: the normal processing time D_i and the total recovery time Θ_i. The time required for recovery is determined by the function $r(X_{i-1})$, which depends on the reprocessing time X_{i-1} associated with failure F_i and a constant time interval needed for rollback. A failure during error recovery interrupts the recovery process and initiates a new recovery. The previous recovery process is not resumed. The time until recovery therefore consists in some interrupted recoveries and finally a successful uninterrupted recovery period.

In previous sections the failure process was assumed a Poisson process and, therefore, needed no further description. Only the checkpointing process was defined. In this section a cycle is the time between failures during normal operation and not the time between checkpoints.

Checkpoints are taken only during normal processing D_i, following some checkpoint strategy which determines the normal operation time between checkpoints. In the i-th cycle $\tau_i(1)$ denotes the time production time until the first checkpoint, $\tau_i(k)$ is the time between the $(k-1)$th and the k-th checkpoint in the i-th cycle. In total $k = 1, 2, \ldots, J_i$ checkpoints are completed successfully in cycle i.

The checkpointing time $C_i(k)$ is the time needed to take the k-th checkpoint with $k = 1, 2, \ldots, J_i$ in cycle i. The process $\{C_i(k), k = 1, 2, \ldots\}$ is stationary and independent of the system state. Furthermore, the checkpointing process is assumed to be independent and identically distributed with mean $C > 0$.

The system state is determined by the checkpointing process. Since failures happen independently of the system state a failure may happen before, during or after the first checkpoint in a cycle. In the first case, the checkpoint interval $\tau_i(1)$ is larger than the normal processing time D_i. The reprocessing time X_i associated with failure F_{i+1} is therefore the sum of the production time D_i, which has not been checkpointed, plus the reprocessing time X_{i-1} associated with F_i, as shown in Fig. 9.13a.

A failure during checkpointing, which is the second case, illustrated in Fig. 9.13b, is assumed equivalent to a failure immediately before checkpointing with respect to the recovery time. The third case is illustrated in Fig. 9.13c. The reprocessing time X_i associated with failure F_{i+1} is the time between completion of the J_i-th checkpoint and the failure. These three cases must be considered in the analysis of the model which is in detail carried out in [146], we will only present the results.

Some more definitions are necessary. Let $V_i(j)$ be the cumulative production time between the end of the error recovery and the j-th checkpoint in cycle i, i.e.

$$V_i(j) = \begin{cases} \sum_{k=1}^{j} \tau_i(k), & j = 1, 2, \ldots, \\ 0 & \text{otherwise}, \end{cases}$$

and let $B_i(j)$ be the sum of the first j checkpointing times during the i-th cycle,

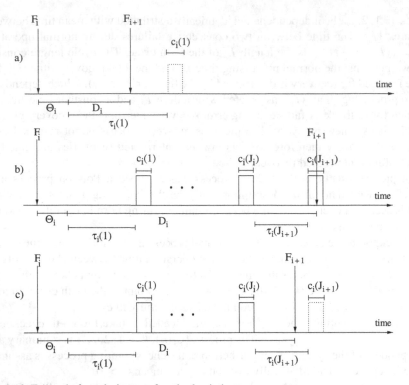

Fig. 9.13 Failure before, during, or after checkpointing

$$B_i(j) = \begin{cases} \sum_{k=1}^{j} C_i(k), & j = 1, 2, \ldots, \\ 0 & \text{otherwise.} \end{cases}$$

Then the starting time $S_i(j)$ and the ending time $E_i(j)$ of the j−th checkpoint during the i−th cycle equal

$$S_i(j) = V_i(j) + B_i(j - 1), \qquad j = 1, 2, \ldots,$$

and

$$E_i(j) = V_i(j) + B_i(j), \qquad j = 1, 2, \ldots.$$

For further analysis the distribution of the reprocessing time must be known. Since failures can happen during reprocessing the time needed for reprocessing can be longer than the actual reprocessing requirement. For a given reprocessing requirement x let Y_x denote the random variable for the time to reprocess work x with pdf $f_{Y_x}(y)$. Let furthermore D_x be the normal processing time associated with the reprocessing time x with PDF F_{D_x}. The unit step function is defined as

$$u(a) = \begin{cases} 1 & \text{if } a \geq 0 \\ 0 & \text{otherwise,} \end{cases}$$

and the conditional function $U(x)$ equals one if its argument is a true condition, i.e.

$$U(\text{COND}) = \begin{cases} 1 & \text{if COND} = \text{TRUE} \\ 0 & \text{otherwise.} \end{cases}$$

In what follows the index i for the considered cycle is omitted for clarity and the equations hold for any arbitrary cycle. The probability density of the reprocessing time can then be expressed as (for a derivation see [146])

$$\begin{aligned}
f_{Y_x,C}(y) = &\; f_{D_x}((y-x) \cdot u(y-x)) \, U(\tau_x(1) > (y-x)) \\
&+ \left((1 - F_{D_x}(S_x(1))) - (1 - F_{D_x}(E_x(1))) \right) U(\tau_x(1) = (y-x)) \\
&+ \sum_{k=2}^{\infty} \Big[f_{D_x}(E_x(k-1) + y) \, U(\tau_x(k) > y) \\
&+ \left((1 - F_{D_x}(S_x(k))) - (1 - F_{D_x}(E_x(k))) \right) U(\tau_x(k) = y) \Big], \quad y \geq 0.
\end{aligned}$$
(9.52)

The dependence on the checkpointing times is eliminated by integrating over the density of the reprocessing time with respect to the checkpointing times

$$f_{Y_x}(y) = \int_0^\infty \int_0^\infty \cdots \int_0^\infty f_{Y_x,C}(y) \, dF_C(c(1)) \, dF_C(c(2)) \ldots \qquad y \geq 0.$$
(9.53)

where $c(k)$ is the checkpointing cost of the k-th checkpoint, as shown in Fig. 9.13. As $i \to \infty$ the stationary pdf of Y, assuming it exists, is the solution ϕ of the equilibrium equation

$$\phi(y) = \int_0^\infty f_{Y_x}(y)\phi(x) \, dx, \qquad y \geq 0,$$
(9.54)

subject to

$$\int_0^\infty \phi(x) \, dx = 1.$$

The stationary PDF then is obtained as usual by integrating over the density

$$\Phi(y) = \int_0^y \phi(x) \, dx.$$

The system availability is defined as the ratio of the mean production time in a cycle $E[N]$ and the mean length of a cycle $E[L]$,

$$A = \frac{E[N]}{E[L]}. \tag{9.55}$$

The denominator is given by

$$E[L] = \frac{1}{\gamma} \int_0^\infty \frac{1}{1 - F_F(r(x))} \, d\Phi(x). \tag{9.56}$$

Recall that γ is the failure rate, $r(x)$ is the recovery time. $F_F(.)$ is the CDF of the interfailure time distribution.

The nominator cannot be expressed as straight forward. It is defined as

$$E[N] = \int_0^\infty E[N_x] \, d\Phi(x), \tag{9.57}$$

where

$$E[N_x] = \int_0^\infty \int_0^\infty \cdots \int_0^\infty E[N_{x,C}] \, dF_C(c(1)) \, dF_C(c(2)) \ldots$$

and

$$E[N_{x,C}] = \frac{1}{1 - F_F(r(x))} \cdot \left(\frac{1}{\gamma} - \int_0^{r(x)} y \, dF_F(y) - \sum_{j=1}^\infty \left((B(j-1) + r(x)) \right. \right.$$

$$\left. \left. [F_F(S_x(j) + r(x)) - F_F(E_X(j-1) + r(x))] + \int_{S_x(j)+r(x)}^{E_X(j)+r(x)} (y - V_x(j)) \, dF_F(y) \right) \right).$$

This completes the specification of the model. In its general form usage of the model is rather laborious. It has, however, some compact special cases, such as Model A in [24] and the same approximation in [181] and Sect. 9.1.1 as well as in Sect. 9.2.1. Another special case is the model in [52] which is discussed in the next section.

As being very general, the model is very interesting from a theoretical point of view. For practical purposes, such as computing the optimal checkpoint interval in a closed-form expression some simplifications are necessary. Let us therefore assume that no failures occur during checkpointing and recovery. The system availability (9.55) then takes a much simpler form. The mean cycle length reduces to the mean interfailure time

$$E[L] = \frac{1}{\gamma}.$$

In the known affine linear equation $r(x) = R + \rho x$ for the recovery time the affine term R is a constant portion for restoring the previous checkpoint and the linear part ρx is the time needed to reprocess the audit trail. Reprocessing time of the work requirement x is determined by the compression factor ρ. Using $E[X]$ for the mean reprocessing time, the production time per cycle in equilibrium can be expressed as

$$E[N] = \frac{1}{\gamma} - R - \rho E[X] - C \cdot \int_0^\infty \sum_{j=1}^\infty \left(1 - F_{F'}(V_x(j))\right) d\Phi(x).$$

Consequently, the system availability is

$$A = 1 - \gamma \left(R + \rho E[X] + C \cdot \int_0^\infty \sum_{j=1}^\infty \left(1 - F_{F'}(V_x(j))\right) d\Phi(x)\right). \qquad (9.58)$$

This equation has an intuitive interpretation. Within one time unit there are on the average γ failures. For each failure the mean total recovery time is given by the first two terms in the parentheses, while the last term expresses the total checkpointing time.

The equidistant checkpoint strategy is given by the sequence of checkpoint intervals, which are

$$\tau_x(j) = \begin{cases} \tau - x, & j = 1 \\ \tau, & j = 2, 3, \ldots. \end{cases}$$

Note that the first checkpoint interval in a cycle starts after recovery and reprocessing is completed, while all other checkpoint intervals are the time between two consecutive checkpoints. In [146] it is shown that equidistant checkpoint intervals imply a uniform distribution of the reprocessing time, $\phi(x)$,

$$\phi(x) = \frac{1}{\tau}, \qquad 0 \le y \le \tau$$

with mean reprocessing time $E[X] = \frac{\tau}{2}$.

The equidistant checkpointing strategy has availability

$$A_\tau = \frac{1 - \gamma(R + \rho\tau/2)}{1 + C/\tau} \qquad (9.59)$$

It is interesting to see that the availability does not depend on the interfailure time distribution, but only on the mean time between failures. Using different interfailure time distributions does therefore not change the system availability.

Differentiating the availability with respect to τ and equating the derivative to zero gives the optimal checkpoint interval, which maximises system availability.

$$\tau^* = \sqrt{C^2 + \frac{2C(1/\gamma - R)}{\rho}} - C. \qquad (9.60)$$

The optimum checkpoint interval is a real number only as long as the expression under the square root is greater than or equal to zero,

$$C^2 + \frac{2C(1/\gamma - R)}{\rho} > 0.$$

This limits all parameters. For the system failure rate, for example, must hold

$$\gamma < \frac{2}{2R - \rho C}.$$

More strictly, the optimal checkpoint interval must be non-negative and therefore a useful result is obtained only when

$$C^2 + \frac{2C(1/\gamma - R)}{\rho} > \sqrt{C}$$

which holds for

$$\gamma < \frac{2}{(C^{-\frac{1}{2}} - C)\rho + 2R}.$$

The bound on γ contains as parameters only the mean time for checkpoint establishment, rollback and recovery. Therefore, the range of the failure rate, for which checkpointing is applicable depends on the checkpoint and recovery costs. This is intuitively convincing.

The curves in Fig. 9.14 compare the optimal checkpoint interval as computed using (9.60) with the one in (9.43), since both maximise availability, but in slightly different models. The parameters are again rollback cost $R = 0.5$, compression

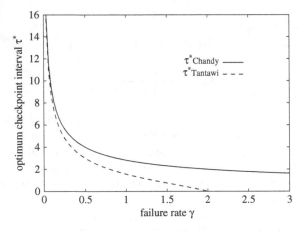

Fig. 9.14 Comparison of the optimal checkpoint interval with an approximation

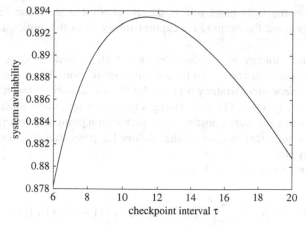

Fig. 9.15 System availability for changing checkpoint interval

factor $\rho = 1/8$ and cost of checkpointing $C = 0.5$. For these parameter values the optimal checkpoint interval τ^* as defined in (9.60) evaluates to 11.34, which is slightly less than 12, the result for (9.43) in Sect. 9.2.1.

For the given parameters only a failure rate of less than 2.0 leads to useful results. The model suggests that if the failure rate is more than 2.0 checkpointing does not help or is rather a burden on system availability.

In Fig. 9.14 the curve displaying (9.60) always shows a smaller optimum checkpoint interval in the model presented in this section, as compared with the model from [24], presented in Sect. 9.2.1. The approximation (9.43) always overestimates the optimal checkpoint interval. The system availability for different checkpoint intervals in Fig. 9.15 shows that it is always better to take a checkpoint interval rather too large than too small. The increase before the maximum of the curve is stronger than the decrease after the maximum.

The equicost checkpointing strategy is defined as the sequence of checkpoint intervals which satisfy

$$\int_{0}^{\tau(j+1)} \rho y \Gamma(V(j) + y)\, dy = C, \qquad j = 1, 2, \ldots, \tag{9.61}$$

where

$$\Gamma(x) = \frac{f_{F'}(x)}{1 - F_{F'}(x)} \qquad x \geq 0$$

is the failure rate and ρ is the compression factor that determines the reprocessing time of the audit trail during recovery. The integrand on the left-hand side is the reprocessing time in case a failure occurs within the $(j + 1)$–th checkpoint interval and on the right-hand side is the checkpoint cost. If the failures occur in a Poisson process the optimal checkpoint intervals in the equicost strategy are equidistant.

The equicost strategy is a *failure-dependent* checkpoint strategy because the checkpoint intervals depend on the failure time distribution. At the same time it is a *reprocessing independent* strategy because the first checkpoint is taken immediately after error recovery, i.e. $\tau_x(1) = 0$. Using a reprocessing independent strategy the moment after error recovery constitutes a regeneration point and no reprocessing of the recovery is necessary in case another failure happens after recovery and before the first checkpoint.

The mean reprocessing time becomes

$$
E[X] = E[N] - \sum_{j=2}^{\infty} \Big(V(j) - V(j-1) \Big) \big(1 - F_{F'}(V(j)) \big). \tag{9.62}
$$

Substituting (9.62) into (9.58) gives

$$
A_C = 1 - \gamma \Big(R + \rho E[N] + C + \sum_{j=2}^{\infty} \big(C - \rho(V(j) - V(j-1)) \big) \big(1 - F_{F'}(V(j)) \big) \Big). \tag{9.63}
$$

Unfortunately, (9.63) depends not only on the mean production time $E[N]$ but also on the full distribution of the times between failures $F_{F'}(.)$. The mean production time can be determined using

$$
E[N] = \frac{A}{\gamma} \tag{9.64}
$$

so that (9.64) and (9.63) may be solved iteratively. As stated in [146] the algorithm typically converges in very few iterations.

The parameters of the previous graphs cannot be used again, since no implementation of the iterative algorithm is at hand and only the data given in [146] is displayed. The parameters used there are the Weibull distribution for the time between failures with mean interfailure time $1/\gamma = 60$ h and shape parameter ranging from 0.2 to 5. The mean checkpoint time is $C = 1$ min, $R = 6$ min and compression factor $\rho = 0.5$.

Figure 9.16 shows that while obviously system availability for the equidistant strategy remains constant over all values of the shape parameter of the Weibull failure time distribution, the equicost strategy achieves the better availability the more pronounced the increasing or decreasing failure rate is. For shape parameter 1, which is the case when the Weibull distribution reduces to the exponential distribution, availability for both strategies is almost identical and, in fact, the equidistant strategy seems slightly better, since then it is the optimal strategy. A Figure in [146] shows the checkpoint intervals for increasing and decreasing failure rate of

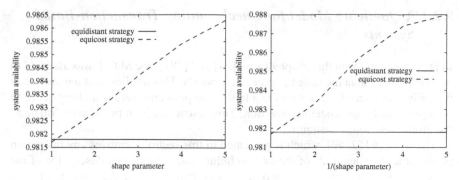

Fig. 9.16 System availability for increasing (*left*) and decreasing (*right*) failure rate

the Weibull distribution. For increasing failure rate the first checkpoint intervals are very long and as more checkpoints are being taken and the failure rate of the system increases the checkpoint intervals become shorter. For decreasing failure rate the opposite holds.

The equicost strategy is significantly superior in performance to the equidistant strategy, in computational complexity it is unfortunately not competitive. Had efficient and fast computation of the optimum checkpoint interval lengths become available over the past decades it would certainly have found its application in practice. The computation of the optimum checkpoint interval for the equicost strategy is complicated by the fact that the length of each checkpoint interval must be determined individually.

Minimisation of checkpoint and recovery cost has also been done for Model A in [24], an example of which is shown in Fig. 9.11. The model used there is a special case of the one treated in this section and it shows that minimising total cost is feasible in restricted cases.

In [92] a model is analysed that allows for failures during checkpointing and recovery. At the same time the checkpointing as well as the recovery cost can be generally distributed. The failure rate of the system is a cyclic function of the time since the most recent checkpoint. The model corresponds to a Markov renewal decision process. The underlying discrete state model has the same three states (failed, operational, checkpointing) as the model in [146], which has been discussed in detail in this section. The decision is recovery in the failed state, and the length of the interval until the next checkpoint in both other states. A reward function measures system availability. This reward function must be optimised with respect to the taken decisions. Optimisation is carried out through a dynamic programming algorithm for which the reward function must be discretised. Solving this model is therefore very costly. The results presented in [92] use the same parameters as the model with the Weibull failure rate in [146] and the results differ by less than 0.05%. This shows that both methods lead to correct results, but also that the additional modelling detail in [92] is not worth while.

9.2.3 A Queueing Model for Checkpointing Transaction-Based Systems

In previous sections in this chapter (as in [24] and [181]) the MTBF was assumed to be much longer than the time between checkpoints. This implies, that the audit trail typically is not very long at the time a failure happens and there is no long queue to be reprocessed after system restoration. This restriction is in practice realistic and therefore no serious constraint.

The work in [52, 53], which is presented in this section, removes any restriction on the order of magnitude of the system failure rate. The formal description of the considered systems consists of two parts, a system model, and a queueing model as shown in Fig. 9.17. The three states *working, recovery* and *checkpointing* and the transitions between them constitute the system model, while the processing of tasks in the *working* state with arrival rate and service time specifies the queueing model. The queueing model explicitly models the transactions in the system and their arrival, waiting and being serviced. None of the earlier models in this chapter did so. The model in this section can be used to compute system availability as well as completion time of transactions.

The stochastic process $X_t, t \geq 0$ is defined as

$$X_t = \begin{cases} 0 & \text{if the system is operational} \\ 1 & \text{if the system is recovering from a failure} \\ 2 & \text{if the system is taking a checkpoint} \end{cases} \qquad (9.65)$$

In the following the dynamics of this stochastic process are defined.

The work in [52, 53] distinguishes between the uptime of a a system and its operational time. The system is up also during checkpointing, but it is not ready to do useful work during checkpointing. The operational time is the time the system spends in state *working*, as shown in Fig. 9.17. The random variable Y with

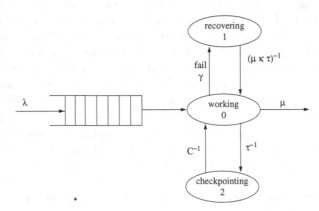

Fig. 9.17 System model

probability distribution function $F(y)$ and density function $f(y)$ denotes the total time in state *working* between two successive visits to state *checkpointing*. The total uptime between checkpoints Y can be interrupted by several failures and the recovery periods thereafter. The expected duration of the total operational time between two checkpoints is

$$E[Y] = \int_0^\infty y \, dF(y).$$

Later, we will see that a deterministic choice of Y optimises system availability.

Checkpointing cost is the time the system spends in state *checkpointing*. This time is random, independent of the past history and has a general distribution function $C(y)$ with finite expectation

$$E[C] = \int_0^\infty y \, dC(y).$$

The system transitions to the failed state (state *recovery*) only from the operational state, not from the checkpointing state. This means that no failures can happen during checkpointing. Failures happen in a Poisson process at rate γ and γ is the failure rate of the server.

The recovery time $r(Y_t)$ is defined in a more complex way. The recovery time is age dependent with respect to the most recent checkpoint. In other words, the time Y_t denotes the operational time since the most recent checkpoint, until system failure. Y_t can be either the time between a checkpoint and the first failure, or it can be the cumulative operational time since the most recent checkpoint where one or more failure and recovery periods have already happened. The recovery periods are then excluded from the total operational time.

Of interest are the stationary state probabilities

$$\Pi_j = \lim_{t \to \infty} \Pr\{X_t = j\}, \qquad j = 1, 2, 3$$

of the regenerative process $(X) = \{X_t, t \geq 0\}$.

The solution is

$$\Pi_0 = \frac{E[Y]}{E[C] + E[Y] + \gamma \int\limits_0^\infty dF(y) \int\limits_0^y r(x) \, dx} \tag{9.66}$$

$$\Pi_1 = \frac{\gamma \int\limits_0^\infty dF(y) \int\limits_0^y r(x) \, dx}{E[C] + E[Y] + \gamma \int\limits_0^\infty dF(y) \int\limits_0^y r(x) \, dx} \tag{9.67}$$

and, of course,

$$\Pi_2 = 1 - \Pi_0 - \Pi_1. \tag{9.68}$$

The optimum checkpoint interval could be either the total operational time Y between successive checkpoints, such that the system availability Π_0 is maximised, or the checkpoint interval could include failure recovery periods that lie between checkpoints. The work in [52] erroneously claims that [24] includes recovery periods into the checkpoint interval. On the contrary, the choice in [52] of counting only the operational time for the checkpoint interval seems to be a common and natural choice also taken by other authors. A contribution of the work [52] is the use of a general distribution $F(y)$ for the checkpoint interval. The optimum checkpoint interval is $F(y)$ such that the availability Π_0 is maximised.

As in the previous sections, again $r(x)$ is defined as an affine linear function

$$r(x) = R + \rho x \qquad R, \rho > 0,$$

where R is the affine term, representing the recovery of e.g. a data base and ρx is the time needed to reprocess the audit trail. The work requirement x is reprocessed as determined by the compression factor ρ. Then

$$H(y) = \int_0^y r(x)\,dx = \frac{R}{2}y^2 + \rho y$$

and

$$E[H] = \frac{R}{2}E\left[y^2\right] + \rho E\left[y\right] \geq H(E[Y]) = \frac{R}{2}E\left[y\right]^2 + \rho E\left[y\right].$$

And the distribution function of the optimum checkpoint interval is

$$F^*(y) = \begin{cases} 1 & \text{if } y \geq \tau \\ 0 & \text{if } y < \tau \end{cases}$$

for fixed $\tau \geq 0$. Note that the formulation is that of an arbitrary checkpoint interval distribution, but the definition corresponds to that of a constant deterministic interval of length τ. The system availability using the optimal checkpoint interval distribution evaluates to

$$\Pi_0(F^*(y)) = \frac{1}{1 + E[C]/\tau + \gamma\rho + R\gamma\tau/2}.$$

To determine the parameters R and ρ the queueing system must be examined. Let us assume transactions arrive at rate λ and are served at rate μ in a first-come-first-served fashion and let interarrival and service times be exponentially distributed.

The probability that the system is operational (in state '0') and idle is denoted $p(0, 0)$.

The stationary probability that the server is idle and operational is

$$p(0, 0) = 1 - \frac{\lambda}{\mu} \Pi_0.$$

Substituting into the system availability using the optimal checkpoint interval $\Pi_0(F^*(y))$ one obtains

$$\Pi_0 \left(1 + \frac{\mathrm{E}[C]}{\tau} + \gamma\rho \right) = 1 - \frac{k\gamma\tau\lambda}{2\mu}. \tag{9.69}$$

Equating the derivative of the availability with respect to the checkpoint interval gives the optimum checkpoint interval

$$\tau^* = \frac{\mathrm{E}[C]}{1 + \rho\gamma} \left(\sqrt{1 + \frac{2(1 + \rho\gamma)}{\lambda/\mu k \mathrm{E}[C]}} - 1 \right). \tag{9.70}$$

Several interesting observations can be made. First, the optimal checkpoint interval, which is the total operational time of the system between two checkpoints, is deterministic and of constant length. Second, the optimum checkpoint interval depends on the reciprocal of the processing load of the system (λ/μ). The higher the load, the shorter the optimum checkpoint interval, which seems reasonable.

Figure 9.18 compares the optimum checkpoint interval using (9.70) and the approximation (9.43) (which for $\rho = 1$ is identical with (9.5) from [181]) as a function of the failure rate γ. The chosen parameters are mainly taken from [52] and they are $\mathrm{E}[C] = 0.5$, $\rho = 0.9$, $\mu = k = 1.0$, λ as indicated in the figure. The parameter k indicates the proportion of transactions that need reprocessing after a failure. For a data base system only writing transactions need reprocessing, reading transactions do not. We assume here that all transactions must be reprocessed. The figure gives some insight in the quality of the approximation as well as in the characteristics of the optimum checkpoint interval obtained with the queueing analysis. The two curves showing the optimum checkpoint interval defined in (9.70) differ in the utilisation of the system. Since the processing rate of the queueing system is set to 1, the transaction arrival rate λ equals the utilisation of the queue. The failure rate in a system with checkpointing usually is below 10^{-2}. The curves have an interesting shape in the range from 0 to 0.5, after that point they can be extrapolated linearly.

It is interesting to see that the simple approximation (τ_{Chandy}) for low system failure rate matches very well with the exact formula for the optimum checkpoint interval with the high system load, while for high failure rate the approximation increasingly overestimates both curves, still being closer to the one for the low system load ($\lambda = 0.65$). Considering its simplicity, the approximation in (9.43) shows remarkably high quality.

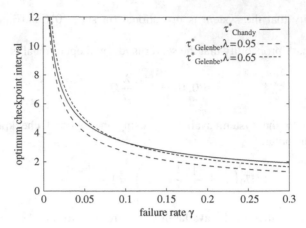

Fig. 9.18 Exact and approximate optimum checkpoint interval ($\rho = 1$)

In [52] a simpler formula for the case of heavily loaded systems is derived. Under heavy load the probability of an idle server in the operational state is almost zero, i.e. $p(0, 0) \approx 0$. and for the fraction in the root of (9.70) holds

$$\frac{2(1 + \rho\gamma)}{\lambda/\mu k \mathrm{E}\,[C]} \gg 1,$$

such that this term dominates the formula for the optimum checkpoint interval and the 1 on either side of the fraction can be ignored. Then the optimum checkpoint interval can be written as

$$\tau^* \approx \sqrt{\frac{2\mathrm{E}\,[C]}{\lambda/\mu\gamma k(1 + \rho\gamma)}}. \tag{9.71}$$

If reprocessing is ignored and therefore $\rho = 0$ and $k = 1$ we have Chandy's known approximation.

In [53], as in [52] the recovery time is deterministic, but age-dependent, where the age y_t is the processing time since the most recent checkpoint excluding recovery of previous failures in the considered checkpoint interval. In [53] the recovery time is defined as $\mu k y_t$, which is the time needed to reprocess all transactions that were processed between the most recent checkpoint and system failure. Then in [53] the system availability $A = \Pi_0$ evaluates to

$$A = \frac{1}{1 + \mu k \tau\gamma + \mathrm{E}\,[C]\,/\tau.} \tag{9.72}$$

It is interesting to see that only the transaction processing rate is a parameter of system availability, not the transaction arrival rate. The system can therefore be in overload, but still available. Additionally, it was shown in [53] that system availability is always less than the utilisation of the queue, i.e.

$$A < \frac{\lambda}{\mu}$$

which is an interesting property, even more, since analytically availability is independent of the utilisation.

Also of importance is the response time W of a transaction, as derived in [53]. The response time of a transaction is the time between the arrival of a transaction to the system and the completion of its processing. This time includes the waiting time in the queue and reprocessing after a system failure. The response time is given by

$$W = \frac{\frac{1}{\mu} + A^2 \left(\gamma (\mu k \tau)^2 + \frac{E[C]^2}{\tau} \right)}{A - \lambda/\mu}. \qquad (9.73)$$

The optimum checkpoint interval τ_W that minimises the transaction response time can be approximated by the checkpoint interval that minimises $\gamma (\mu k \tau)^2 + E[C]^2/\tau$, the decisive term in (9.73), as derived in [53]

$$\tau_W = \sqrt[3]{\frac{E[C]^2}{2\gamma (\mu k)^2}}. \qquad (9.74)$$

Figure 9.19 shows different curves with their minima at different values of the checkpoint interval for different transaction arrival rate. The fact that the transaction arrival rate is no factor in the formula computing the optimum checkpoint interval identifies τ_W as an approximation.

Figure 9.19 shows the system availability and the transaction response time for different values of the checkpoint interval length. As pointed out in [53] in transaction processing systems not only the system availability but also the transaction response time is of interest. The curves for the transaction response time in Fig. 9.19 are scaled by a factor 10^{-1} for better visualisation of all three curves together. The chosen parameters are mostly the same as those used for Fig. 9.18, $\gamma = 10^{-3}, \mu = 1.0, k = 1.0, E[C] = 0.5$. The arrival rate of transactions is 0.2 and 0.5 as indicated in the plot. Since the service rate $\mu = 1$ the transaction arrival rate equals the utilisation of the transaction processing system. In normal operation the utilisation of such systems is quite low. As shown in [52] already for a utilisation of 0.65 the queue length explodes, even though system availability is still more than 0.7. Also the transaction response time increases as the utilisation of the system increases. As suspected, the optimum checkpoint interval that maximises system availability is significantly different from the checkpoint interval length that minimises transaction response time. Furthermore, the checkpoint interval that minimises transaction response time is not the same for different system utilisation. It slightly increases for increasing load, as does of course the expected transaction response time. As the load on the system grows, checkpointing should be done slightly less frequently. This is because the checkpointing imposes additional load

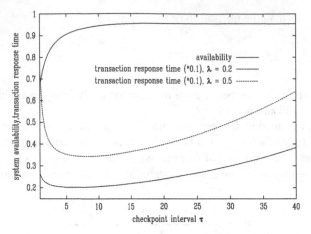

Fig. 9.19 Optimum checkpoint interval that maximises availability and transaction response time

on the system. The more the system is loaded, the less extra checkpointing cost it can take when minimising the transactions' response time.

For the parameters used in the plots in Fig. 9.19 the optimum checkpoint interval to minimise transaction response time, as given in (9.74), is $\tau_W = 6.3$. for both values of λ. The curve using the higher load, $\lambda = 0.5$ clearly has a slightly larger optimum checkpoint interval.

Advanced queueing analysis is applied in [9] in order to compute also the expected waiting time of transactions and expected number of transactions in the system.

In [54] the model is extended to use an age dependent failure rate $\gamma(y)$, which is a natural choice and can be implemented using the Weibull distribution. Unfortunately, no closed-form expression for the optimum checkpoint interval exists, which must then be computed numerically. Numerical solutions are less attractive in practice.

In [90] the task length, or work requirement of a task, is also considered, furthermore, idle periods, first tasks after an idle period and busy periods are treated separately. If tasks arrive during a busy period of the system in a Poisson process at rate λ and the total recovery time of the system is denoted by the generally distributed random variable R, then the expected completion time of the deterministic work requirement w can be expressed as

$$E\left[T(w)\right] = a \cdot \left(\frac{1}{\tau} + \gamma\right) w + a \cdot \ln\left(\frac{1/\tau + \gamma e^{(1/\tau + \gamma)w}}{1/\tau + \gamma}\right), \qquad (9.75)$$

where

$$a = \tau\left(1 + \frac{E[C]}{\tau} + \gamma E[R]\right).$$

For long tasks the model becomes similar to the completion time models of program level checkpointing [113] in Sect. 9.1.1. As w grows large one can use the limit

$$\lim_{w \to \infty} \left(\frac{1/\tau + \gamma e^{(1/\tau + \gamma)w}}{1/\tau + \gamma} \right) = \frac{1}{1 + \gamma \tau}$$

as an approximation and the expected completion time in a system with checkpointing is a linear function of the work requirement, i.e.

$$E[T(w)] \approx a \cdot \left(\frac{1}{\tau} + \gamma \right) w - a \cdot \ln(1 + \gamma \tau). \tag{9.76}$$

Figure 9.20 plots (9.76) for different parameter values. The constant parameters are the task length $w = 300$, the expected repair time $E[R] = 25$ and the expected checkpoint cost $E[C] = 0.5$. As usual a system with the low failure rate has a much longer optimal checkpoint interval and most tasks do not require much more time than their work requirement. This changes abruptly as the failure rate increases. The checkpoint interval becomes shorter and the expected task completion time rapidly increases if a too long checkpoint interval is chosen.

In [90] also an expression for the variance of task completion time as well as limiting cases are given. Checkpointing strategies are evaluated by means of the *stretch factor*, the relative overhead imposed by checkpointing,

$$\mu_\tau(w) = \frac{E[T(w)]}{w},$$

where $\tau > 0$. Checkpointing is beneficial for those values of τ where $\mu_\tau(w) < \mu_0(w)$. Unfortunately, deriving conditions under which this inequality holds is very

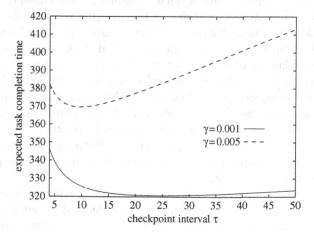

Fig. 9.20 Optimum checkpoint interval with respect to the system failure rate

laborious [90]. An approximation for the optimum checkpoint interval which minimises the stretch factor is

$$\tau^* = \frac{\sqrt{\mathrm{E}\,[C]}}{\sqrt{\gamma(1 + \gamma\mathrm{E}\,[R])}} \tag{9.77}$$

with corresponding lower bound on the stretch factor itself

$$\inf_\tau \mu_\tau(w) = \left(\sqrt{1 + \gamma\mathrm{E}\,[R]} + \sqrt{\gamma\mathrm{E}\,[C]}\right)^2. \tag{9.78}$$

In [90] these approximations are stated to provide good results, which cannot be
easily verified. Queueing analysis in [90] leads to similar results for the expected
task completion time and the optimum checkpoint interval.

An additional checkpoint just after recovery is placed in [36].

9.3 A Trade-Off Metric for Optimal Checkpoint Selection

We have seen in the previous sections that for program level checkpointing a typical
metric is the expected completion time of a task, or its distribution. When evaluating
system level checkpointing strategies usually the system availability is the metric of
interest. Using these metrics optimal checkpoint intervals are determined. In [146]
the equicost strategy is used to determine optimal checkpoint interval length. The
equicost strategy optimises not only system availability but also its production time,
which is the availability relative to the system failure rate. Note that the equicost
strategy yields an improvement only if the times between failures are other than
exponentially distributed.

In [84] both the expected task completion time and the system availability are
criticised as being not appropriate. When looking at their computation in detail, both
metrics average over all tasks and time. Checkpointing increases the completion
time of all tasks, where tasks that complete without failure perceive degraded service
for the benefit of failed tasks. With checkpointing failed tasks have shorter recovery
time. But since relatively only very few tasks experience a system failure very many
tasks unnecessarily have prolonged completion time. In the opinion of the authors of
[84] simply optimising completion time or availability does not optimise the trade-
off between the price non-failed tasks pay and the benefit failed tasks receive. This
statement is not obviously correct and not proven in the paper.

A trade-off ratio can be defined for program-level checkpointing. In [84] program-
level checkpointing is applied to *general purpose systems* and the completion time
of tasks is the metric of interest.

The idea in [84] is to divide the tasks into those that fail and those that do not
fail. Then the expected completion time for non-failed tasks when using n checkpoints $\mathrm{E}\,[T_0(n)]$ and the expected completion time of tasks that experience failure and use n checkpoints $\mathrm{E}\,[T_f(n)]$ is computed individually. The trade-off ratio

$TR(n)$ relates the gain in both performance metrics when increasing the number Of checkpoints.

$$TR(n) = \frac{E\left[T_f\right](n-1) - E\left[T_f(n)\right]}{E\left[T_0(n)\right] - E\left[T_0(n-1)\right]}. \qquad (9.79)$$

This ratio is positive as long as the expected completion time of the failed tasks decreases with more checkpoints while the expected completion time of the tasks without failures increases when using more checkpoints. It is worth realising that when the expected completion time of both groups is weighed with the number of non-failed and the number of failed tasks respectively then the overall completion time is obtained. In [84] a failure rate of the order of at most 10^{-3} per hour is assumed. But the trade-off ratio gives equal importance to the one failed task and to the large number of non-failed tasks.

Figure 9.21 uses data computed for a system with a failure rate of 10^{-4} and checkpointing cost of 0.5 ms.

Figure 9.21 shows the expected completion time $E\left[T_f\right]$ of failed tasks for two different task lengths and increasing number of checkpoints, the expected completion time of non-failed tasks with two different task lengths over increasing number of checkpoints and the trade-off ratio for both task lengths. The trade-off ratio compares the change in the completion time of a failed task when adding a checkpoint with the increase in completion time of non-failed tasks when adding a checkpoint. Consequently, a large positive ratio indicates that either the failed tasks considerably reduce their expected completion time with adding a checkpoint, or the non-failed tasks experience very little overhead through an additional checkpoint, or both. Desirable is, therefore, a large trade-off ratio.

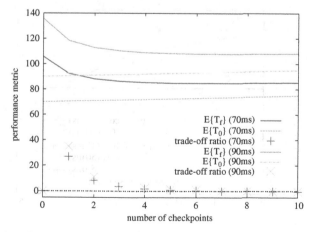

Fig. 9.21 Number of checkpoints versus performance metrics

The trade-off ratio shown in Fig. 9.21 favours no checkpointing at all, while the expected completion time of failed tasks is minimal when using 6 or 7 checkpoints. In [84] these certainly interesting results are merely listed.

For real-time control systems the important metric is the probability of completion before the deadline, since typically completion after the deadline is of no use anymore. In [84, 85] the probability of a dynamic failure $p_{dyn}(n)$ denotes the probability that a task misses its hard deadline when using n checkpoints. The checkpointing cost on all tasks extends their completion time. If $f(t, n)$ denotes the response time distribution of a control task using n checkpoints and $c(t, n)$ the distribution of the checkpointing cost for n checkpoints associated to the response time t of the control task then the mean cost $E[C_n]$ can be expressed as

$$E[C_n] = \int_0^\infty f(t, n)c(t, n)\, dt.$$

The trade-off ratio $(TR(n))$ relates the improvement (be it positive or negative) of the mean cost and the probability of a dynamic failure when increasing the number of checkpoints by one as

$$TR(n) = \frac{p_{dyn}(n-1) - p_{dyn}(n)}{E[C_n] - E[C_{n-1}]}. \tag{9.80}$$

Figure 9.22 shows the probability of dynamic failure p_{dyn}, the mean cost of task completion $E[C]$ as well as their ratio, the trade-off ratio, for a task length of 40 and 50 ms, respectively, computed for a number of checkpoints ranging from 0 to 5. The data is taken from Table 1 in [84].

The considered system has a MTBF of 10,000 h and failures occur in a Poisson process. The checkpointing cost is 0.1 ms, the time for rollback and restart are

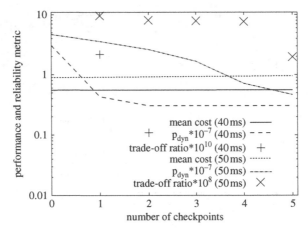

Fig. 9.22 Number of checkpoints versus metrics

each assumed to take 2.0 ms. The points of the trade-off ratio are only shown in Fig. 9.22 where its value is greater than zero, meaning that an additional checkpoint reduces the probability of a dynamic failure. If the probability of dynamic failures remains constant when increasing the number of checkpoints, then the denominator in Eq. (9.80) evaluates to zero and the trade-off ratio equals zero.

Checkpointing in a system with the given parameters is beneficial only for tasks that are not too short. Figure 9.22 illustrates the computed metrics for a task of 40 ms and one of 50 ms. For zero checkpoints the mean cost of a task is already higher than its execution time, due to potential failures of the system. With an increasing number of checkpoints the mean cost of the task of 40 ms hardly increases at all, while the mean cost of the task of length 50 ms increases slightly. The probability of missing the hard deadline decreases as the number of checkpoints increases. As expected, the longer task has a higher probability of missing the deadline than the shorter task and it benefits more from checkpointing, i.e. it experiences a more pronounced decrease of the probability of dynamic failures with increasing number of checkpoints.

One should select the number of checkpoints with a high trade-off ratio. A high trade-off ratio for i checkpoints means that adding the i-th checkpoint reduces the probability of dynamic failures while not increasing the expected cost dramatically. In the example for a task of 40 ms only one checkpoint should be taken, while for a task of 50 ms one should use 4 checkpoints, as for five checkpoints the trade-off ratio abruptly drops. This conclusion was not drawn in [84] and it is not clear how the optimal number of checkpoints can be determined automatically. The authors point out in [84] that considering purely the expected completion time in the given scenario no checkpoints would be used at all.

The issue of appropriate performance and reliability metrics for fault-tolerance mechanisms is certainly very important. Therefore, existing work has been presented in a full subsection. The work in [84], however, is by no means exhaustive or conclusive. But it is an interesting starting point for further work in the analysis of fault-tolerant systems.

9.4 Summary

This chapter has given an overview of existing stochastic models for checkpointing. The purpose of the models is to determine optimal checkpoint intervals, with respect to maximum system availability, minimum task completion time or minimal total recovery and checkpointing costs.

We have seen that the considered metric plays an important role for the choice of the best checkpoint strategy, and we have also seen that a very simple deterministic approximation in most cases gives reasonably good results at very low cost. Very detailed models with high flexibility often require highly complex solution procedures, while the solution as such differs little from the simple approximations.

Checkpointing models were divided into program level checkpointing, which is concerned with long-running tasks and aims at minimising their expected completion time and system level checkpointing, where tasks are typically very short and system availability is more important than expected completion time. Interestingly, for both types of checkpointing the same approximate checkpointing strategy has been derived [181, 24]. Recently, a stochastic activity network (SAN) model has been formulated for coordinated checkpointing of large-scale computing systems [172], taking into account potential failures during checkpointing and recovery operations. It turned out that the effects of correlated failures are considerable and must be taken into account. Large-scale parallel computers are an application with very specific characteristics for which checkpointing is highly beneficial and has therefore undergone some initial analysis [32, 33, 172].

Until recently, the practical applicability of the models, however, has mostly not been an issue in the context of stochastic models. The algorithms to compute the optimum checkpoint interval must perform in real time. All models use simplifications and abstractions of the real system and only the implementation and test of checkpointing strategies can finally show whether those are valid abstractions and whether the models lead to better checkpointing strategies than the equidistant strategy with checkpoint intervals of 11 min IBM uses today [110, 109].

Chapter 10
Summary, Conclusion and Outlook

This book has been concerned with three different, but closely related, mechanisms of software dependability. In this last chapter we will summarise, as well as highlight differences and similarities of the three mechanisms.

The first method was restart, where repeatedly a task is stopped and then restarted from its beginning. This method can be used to reduce task completion time but also to improve the probability of task completion before a deadline. Stochastic models have been used to determine the optimal restart time after which a task is aborted and restarted.

The second method was software rejuvenation, a method of preventive maintenance, which is not concerned with individual tasks, but rather with a whole system. Software rejuvenation requires all tasks on the system to be stopped such that the underlying system software can be rebooted thus solving all memory leakage, buffer overflow and related problems. After a reboot of the system software all processes are restarted and new transactions can be processed. Stochastic models are used to determine the rejuvenation intervals such that a system crash is avoided, or delayed as far as possible. The target function that is optimised is a cost function which expresses the trade-off between rejuvenation costs and costs due to system failure.

The third method was checkpointing, which is the most complex dependability mechanism out of the three. For checkpointing the system state is saved at intermediate points in time and when a failure occurs the affected tasks do not have to restart from beginning, but can roll back to the most recent checkpoint. The system reloads the most recent checkpoint and continues operation from there. The tasks that have been processed since the most recent checkpoint until the system failure are logged in an audit trail, which has to be reprocessed at system recovery. Checkpointing aims at reducing the amount of work lost with a system failure. Checkpoint intervals optimise different criteria: often system availability is considered as the metric of interest, but frequently also transaction response time is the chosen metric. One can, furthermore, use a cost function which formulates the trade-off between the cost of checkpointing and rollback recovery and the cost of failure and restoration after a failure.

K. Wolter, *Stochastic Models for Fault Tolerance*,
DOI 10.1007/978-3-642-11257-7_10, © Springer-Verlag Berlin Heidelberg 2010

Of the three mechanisms checkpointing usually has the most complex system model and the largest variety of considered metrics. Restart has the simplest system model and can optimise only metrics that are functions of the task completion time.

We will now point out the differences and similarities of the three different methods. A comparison is given in Tables 10.1 and 10.2. Even though at first sight the three mechanisms seem to be almost identical, or special cases of one another, on closer inspection there exist fundamental differences. Let us first examine the similarities.

Software rejuvenation and restart both reprocess the halted tasks from the beginning. If checkpointing uses zero checkpoints and upon a failure the task is restarted from the beginning, then checkpointing becomes equivalent to restart and rejuvenation. This is true only if, in addition, failures represent restarts. Restart and rejuvenation both intentionally stop and restart tasks, while normally in checkpointing systems tasks are aborted and recovery is triggered only by system failures. Checkpointing systems stop processing intentionally to take checkpoints, which has no equivalent in neither restart nor rejuvenation. The equivalence of restart with a special case of checkpointing becomes apparent, for example, in Theorem 4.1 on p. 56, which formulates a computational scheme for the moments of completion time. The scheme in fact is a special case of the checkpointing results in [113] (with zero checkpoints and with failures representing restarts).

We have seen that restart, rejuvenation and checkpointing are similar in that all three optimise the first moment of a metric by constant interval lengths. Only for restart higher moments of completion time have been optimised. For higher moments it turned out that constant restart interval length does not achieve the

Table 10.1 Comparison of results for restart, rejuvenation and checkpointing

	Preventive	Reactive	Models	Metrics
Restart	No	Yes	Closed-form	Moments of compl. time
				prob. of meeting a deadline
Rejuvenation	Yes	No	Markovian	Downtime
				Downtime cost
			Non-Markovian	Availability
			Non-Markovian, PN	Expected loss
Checkpointing	Yes	Yes	Analytical, queueing	Availability
			Markovian, non-Markovian	
			closed-form (LST)	Distr. of completion time
			analytical	expected completion time
			analytical	prob. of task completion
				cost

Table 10.2 Comparison of results for restart, rejuvenation and checkpointing

	Optimisation parameter	Recovery
Restart	restart interval	yes
Rejuvenation	Rejuvenation interval	No
Checkpointing	Checkpoint interval	Yes

optimum metric. It would be interesting to see whether for rejuvenation and check-pointing also variable interval lengths are necessary to optimise higher moments of the considered metrics.

Restart is a method which should be applied in situations where the educated guess assumes a transient failure of the system and loss of the request or task. Even though failures are not explicitly modelled, they are implicitly accounted for as very long completion or response times. The assumption is that if a task does not finish within some time, then it very likely has failed and will never finish and hence, it should be restarted. Assuming a transient failure, the next trial is likely to be successful. Restart is about the one task out of very many that fails. The mechanism is very similar to software rejuvenation, only that the latter operates on the whole system, while the former only affects individual tasks. However, they differ in the optimisation criteria. Restart is a mechanism to avoid the few catastrophic cases at the expense of degraded performance of some tasks that perform reasonably well without restart. But in contrast to the common fear restart does not significantly increase the system load. If the restart interval is chosen well, not too many tasks are aborted just before completion. Their completion time is prolonged unnecessarily but at the same time the failed tasks are caught, whose completion time then is enormously reduced.

Rejuvenation also punishes the tasks that are processed when rejuvenation is issued. This is never an issue in the models, because after rejuvenation all tasks are processed faster.

Let us now turn to the differences. The most fundamental difference between restart and both other mechanisms is that the restart model does not allow for failures, while both, rejuvenation and checkpointing do so. As a consequence, restart is not concerned with the aging of a system, while rejuvenation and checkpointing both are. In the restart model task completion time provides an upper bound and restart must only be applied if it can (at least on the long-term average) reduce task completion time. Restart is a mechanism aimed at reducing the considered metric, while rejuvenation rather has the dual purpose. Rejuvenation aims at increasing system life time. Therefore, the acceptable cost of rejuvenation is bounded by the cost of system failure, while the cost of restart is subject to a much tighter bound. Only if the restart cost plus the second attempt together are less than the completion time of the first attempt does restart on an individual task pay off. This is a very restrictive condition and restart is beneficial only for certain completion time distributions and very low restart cost.

In the same direction points a formal difference [166]. While rejuvenation is appropriate in first place for systems with aging behaviour, which are described by a probability distribution with increasing failure rate, restart should be applied only for systems described by probability distributions with decreasing hazard rate. Checkpointing can be applied successfully in both cases.

When considering their behaviour as concerned with system failures the three mechanisms are inherently different. Checkpoint recovery is the only immediately reactive method. Checkpointing is used to react to failures and to mitigate the effects of system failures on processing performance. Checkpointing can also be considered a fault-tolerance mechanism. However, taking a checkpoint is a preventive action. Therefore, checkpointing has a preventive as well as a reactive component. As opposed to this, rejuvenation only aims at acting before a failure happens and hence is exclusively a proactive method to handle failures. This is not precise, as rejuvenation does not include failure treatment. It aims at creating a fault-free system through maintenance. Restart, being the third mechanism, does not explicitly treat system failures, other than that a failure could be reason for long task completion time. Restart purely aims at improving system performance without any direct failure detection or prediction.

For the restart model computationally attractive expressions and algorithms have been developed and the restart model is being applied in practical experiments [128, 129]. This leads us to the open issues. It would be very interesting to implement the algorithm in [152] and carry out experiments. Some detailed questions are still open, such as a when are two checkpoint selections with equal number of checkpoints the same. This algorithm is of particular interest since it seems to be very similar to the algorithms developed for restart. Checkpointing always seems to be a very expensive and therefore practically rarely attractive fault-tolerance method. This issue could be tackled by using the algorithm in [152]. Throughout the text many research questions have been pointed out as those are apparent in their respective context.

In systems where failures can be predicted reliably rejuvenation or checkpointing could be combined with a prediction method. Analysis of possible combinations would be necessary to decide whether the combined approach allows for less frequent rejuvenations and checkpoints and is therefore attractive in computational complexity.

A more general research topic is the solution of timeout problems. As we have seen in Chapter 3 restart can be applied in different contexts. If we generalise further and look at the problem of how to set a timeout then we have a question that arises in almost every area of computing systems. In this book we have seen how restart, rejuvenation and checkpointing relate and that all three are variations and extensions of the plain timeout problem. The most challenging question that remains open is to what extent and in which way the available results for restart, rejuvenation and checkpointing can be applied to the general timeout problem and whether a generic solution and an algorithm can be found. We still study further the fields of Internet services and communication networks and hope to come up with answers and solutions in the near future.

Appendix A
Properties in Discrete Systems

A.1 Cumulative First Moment

Let f be the probability function of a discrete random variable and $F(t_n) = \sum_{i=1}^{n} f(t_i)$ its cumulative probability distribution function, then we prove the following theorem

Theorem A.1

$$\sum_{i=1}^{n} t_i f(t_i) = t_n F(t_n) - \sum_{i=1}^{n-1} F(t_i). \tag{A.1}$$

Without loss of generality we assume that in a discrete system $t_i = i$ and we must prove

$$\sum_{i=1}^{n} i f(i) = n F(n) - \sum_{i=1}^{n-1} F(i). \tag{A.2}$$

Proof Proof by induction. For the base case $n = 1$ the proposition is true since

$$1 \cdot f(1) = 1 \cdot F(1). \tag{A.3}$$

Assume the proposition is true for arbitrary n, then the induction step shows it is also true for $n + 1$

K. Wolter, *Stochastic Models for Fault Tolerance*,
DOI 10.1007/978-3-642-11257-7, © Springer-Verlag Berlin Heidelberg 2010

$$\sum_{i=1}^{n+1} i f(i) = \sum_{i=1}^{n} i f(i) + (n+1) f(n+1)$$

$$= n F(n) - \sum_{i=1}^{n-1} F(i) + (n+1) f(n+1)$$

$$= (n+1) \cdot \sum_{i=1}^{n} f(i) - \sum_{i=1}^{n} f(i) + (n+1) f(n+1) - \sum_{i=1}^{n-1} F(i)$$

$$= (n+1) \cdot \sum_{i=1}^{n+1} f(i) - \left(\sum_{i=1}^{n} \left(f(i) + \sum_{j=1}^{i} f(j) \right) - \sum_{j=1}^{n} f(j) \right)$$

$$= (n+1) \cdot \sum_{i=1}^{n+1} f(i) - \sum_{i=1}^{n} \sum_{j=1}^{i} f(j) = (n+1) \cdot F(n+1) - \sum_{i=1}^{n} F(i).$$

A.2 The Gamma Function

The Gamma function is defined as

$$\Gamma(x) = \int_0^\infty t^{x-1} e^{-t} \, dt. \tag{A.4}$$

It is defined on the real numbers except for $0, -1, -2, \ldots$. For positive natural number the Gamma function can be expressed as a factorial

$$\Gamma(n) = (n-1)!$$

The incomplete Gamma function is defined as

$$\Gamma(x, y) = \int_y^\infty t^{x-1} e^{-t} \, dt. \tag{A.5}$$

Appendix B
Important Probability Distributions

B.1 Discrete Probability Distributions

B.1.1 The Binomial Distribution

A random variable X is said to follow the binomial distribution with parameters n and p (i.e. $X \sim B(n, p)$,) if the probability function of X is given by

$$Pr(X = j) = \binom{n}{j} p^j (1 - p)^{n-j}, \qquad j = 1, 2, \ldots n. \qquad (B.1)$$

In this formula $Pr(X = j)$ is the number of experiments where the event j occurs if a total of n experiments are carried out and the probability of j in each experiment equals p. Expectation and variance of the binomial distribution are

$$E(X) = np \qquad V(X) = np(1 - p).$$

B.1.2 The Multinomial Distribution

Assume we observe N events with n possible outcomes, each happening with probability p_i. The probability distributing th N trials over n possible outcomes is given by the multinomial distribution.

$$P(X_1 = j_1, X_2 = j_2, \ldots, X_n = j_n) = \frac{N!}{j_1! \cdot j_2! \cdots j_n!} p_1^{j_1} \cdot p_2^{j_2} \cdots p_n^{j_n} \qquad (B.2)$$

For $n = 2$ we obtain the binomial distribution.
   Expectation and variance of X_i are

$$E(X_i) = Np_i \qquad V(X_i) = Np_i(1 - p_i).$$

B.1.3 The Geometric Distribution

Let X be a random variable describing the number of trials up to the first success, where each experiment has success probability p. If the n-th trial is the first success, there have been (n-1) unsuccessful trials before.

$$Pr(X = n) = p \cdot (1 - p)^{(n-1)}, \qquad n = 0, 1, 2, \dots \qquad \text{(B.3)}$$

$$E[X] = \frac{1-p}{p}$$

$$Var[X] = \frac{1-p}{p^2}$$

B.1.4 The Poisson Distribution

The Poisson distribution describes the probability of observing k events in the time interval $(0, t]$.

$$P(X = k) = e^{-\lambda t} \frac{(\lambda t)^k}{k!} \qquad x = 0, 1, 2, \dots, \qquad \lambda > 0 \qquad \text{(B.4)}$$

Expectation and variance of the Poisson distribution are equal, they are

$$E(X) = V(X) = \lambda$$

Figure B.1 shows the Poisson probability distribution for two different event rates ($\lambda = 2$ and $\lambda = 4$). Please note that this is a discrete distribution and the points are connected by lines only for better visualisation.

B.2 Continuous Probability Distributions

B.2.1 The Exponential Distribution

The exponential distribution is widely used to describe times between events. Then the cumulative distribution function (CDF) F indicates the probability of the random variable having a value less or equal the observed time x.

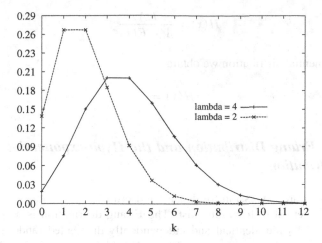

Fig. B.1 Probability distribution function of the Poisson distribution

$$F(x) = Pr(X \le x) = \begin{cases} 1 - e^{-\lambda x} & x \ge 0 \\ 0 & x < 0 \end{cases} \tag{B.5}$$

The exponential distribution has one parameter λ and the mean time between two events is $1/\lambda$.

$$f(x) = \frac{dF(x)}{dx} = \lambda e^{-\lambda x} \tag{B.6}$$

Expectation and variance of X are

$$E(X) = \frac{1}{\lambda} \qquad V(X) = \frac{1}{\lambda^2}.$$

The hazard rate is defined as

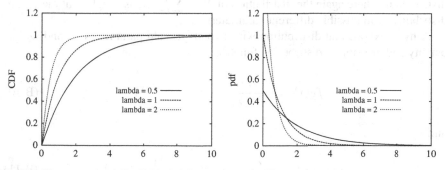

Fig. B.2 The exponential distribution. CDF on the *left*, density on the *right*

$$H(x) = \frac{f(x)}{1 - F(x)}.$$

For the exponential distribution we obtain

$$H(x) = \lambda$$

B.2.2 The Erlang Distribution and the Hypo-exponential Distribution

No straight forward expression for density and probability distribution function of the hypo-exponential distribution exist. The Erlang distribution is a special case where $X_1, \ldots, X_n$ are identical and independently distributed random variables. Then $X = \sum_{i=1}^{n} X_i$ has an n-stage Erlang distribution. For $x, \lambda > 0$ and $n = 1, 2, \ldots$ we have

$$f_X(x) = \frac{\lambda(\lambda x)^{n-1}}{(n-1)!} e^{-\lambda x} \tag{B.7}$$

and

$$F_X(x) = 1 - e^{-\lambda x} \sum_{j=0}^{n-1} \frac{(\lambda x)^j}{j!}. \tag{B.8}$$

The hazard rate of the Erlang distribution is

$$h_X(x) = \frac{\lambda^n x^{n-1}}{(n-1)!} \sum_{j=0}^{n-1} \frac{(\lambda x)^j}{j!}. \tag{B.9}$$

The hypo-exponential distribution is obtained as a generalisation of the Erlang distribution, where again the distribution of $X = \sum_{i=1}^{n} X_i$ is sought, but the X_i are distributed with possibly different parameters λ_i.

A hypo-exponential distribution with two stages and parameters λ_1 and λ_2 has density and probability distribution function

$$f_X(x) = \frac{\lambda_1 \lambda_2}{\lambda_2 - \lambda_1} \left(e^{-\lambda_1 x} - e^{-\lambda_2 x} \right), \tag{B.10}$$

and

$$F_X(x) = 1 - \frac{\lambda_2}{\lambda_2 - \lambda_1} e^{-\lambda_1 x} + \frac{\lambda_1}{\lambda_2 - \lambda_1} e^{-\lambda_2 x}. \tag{B.11}$$

The hazard rate of the hypo-exponential distribution is

$$f_X(x) = \frac{\lambda_1 \lambda_2 \left(e^{-\lambda_1 x} - e^{-\lambda_2 x}\right)}{\lambda_2 e^{-\lambda_1 x} - \lambda_1 e^{-\lambda_2 x}} \tag{B.12}$$

B.2.3 The Hyperexponential Distribution

Let $X_1, \ldots, X_n$ be mutually independent exponentially distributed random variables with parameters $\lambda_1, \ldots, \lambda_n$. Let X be a random variable that is distributed as X_i with probability p_i and $\sum_{i=1}^{n} p_i = 1$. Then X has a n-stage hyperexponential distribution

$$f_X(x) = \sum_{i=1}^{n} p_i \lambda_i e^{-\lambda_i x}, \tag{B.13}$$

and

$$F_X(x) = \sum_{i=1}^{n} p_i (1 - e^{-\lambda_i x}) \qquad x \geq 0. \tag{B.14}$$

With hazard rate function

$$h_X(x) = \frac{\sum_{i=1}^{n} p_i \lambda_i e^{-\lambda_i x}}{\sum_{i=1}^{n} p_i e^{-\lambda_i x}} \qquad x > 0. \tag{B.15}$$

The hyperexponential distribution has higher variance than the exponential distribution. It can be used to model a mixture of different classes of jobs that are distributed with different parameters.

B.2.4 The Mixed Hyper/Hypo-exponential Distribution

The mixed hyper/hypo-exponential distribution can be made to be never, always or sometimes amenable to restart, depending on the chosen parameters. The mixed hyper/hypo-exponential random variable takes with probability p_i a value from an Erlang distribution with N_i phases and parameter $\lambda_i > 0$, $i = 1, 2, \ldots, M$, and $\sum_{i=1}^{M} p_i = 1$. So, we get for the distribution F_M and density f_M (refer, for instance, to [69]):

$$F_M(t) = \sum_{i=1}^{M} p_i \left(1 - \sum_{j=0}^{N_i-1} \frac{(\lambda_i t)^j}{j!} e^{-\lambda_i t}\right),$$

$$f_M(t) = \sum_{i=1}^{M} p_i \lambda_i^{N_i} \frac{t^{N_i-1}}{(N_i-1)!} e^{-\lambda_i t}.$$

In Sect. 4.2 we apply the following parameter values: $M = 2$, with $p_1 = 0.9$, $p_2 = 0.1$; $N_1 = N_2 = 2$, with $\lambda_1 = 20$, $\lambda_2 = 2$; and $c = 0$, unless otherwise stated. This mixed distribution has neither monotonically increasing or decreasing hazard rate, see Fig. 4.6, which implies that it depends on the chosen restart time whether restart improves completion time.

B.2.5 The Weibull Distribution

The Weibull distribution is used to model fatigue failure of components. Its density, probability distribution and hazard rate function are respectively

$$f_X(x) = \lambda^\alpha \alpha x^{\alpha-1} e^{-(\lambda x)^\alpha}, \tag{B.16}$$

and

$$F_X(x) = 1 - e^{-(\lambda x)^\alpha} \qquad x \geq 0. \tag{B.17}$$

With hazard rate function

$$h_X(x) = \lambda^\alpha \alpha x^{\alpha-1} \tag{B.18}$$

and cumulative hazard rate function

$$H_X(x) = \lambda^\alpha x^\alpha. \tag{B.19}$$

In all functions $x \geq 0$ and $\alpha, \lambda > 0$. The hazard rate function of the Weibull distribution can be either increasing, decreasing or constant, depending on the value of α, which makes it a very flexible distribution in describing life times. For $\alpha = 1$ it reduces to the exponential distribution, $\alpha > 1$ implies increasing failure rate (aging), while $\alpha < 1$ means decreasing failure rate. The expectation of the Weibull distribution is

$$E[X] = \frac{1}{\lambda} \cdot \Gamma\left(1 + \frac{1}{\alpha}\right) \tag{B.20}$$

and its Variance

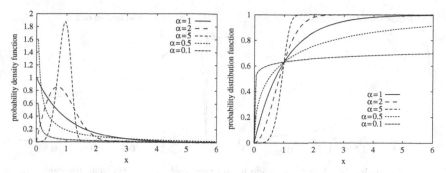

Fig. B.3 The probability density (*left*) and probability distribution function of the Weibull distribution (*right*) ($\lambda = 1$)

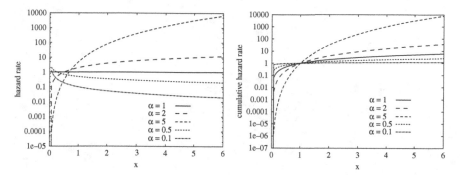

Fig. B.4 The hazard rate (*left*) and cumulative hazard rate function of the Weibull distribution (*right*) ($\lambda = 1$)

$$Var[X] = \frac{1}{\lambda^2} \cdot \left(\Gamma(1 + \frac{1}{\alpha}) + \Gamma(1 + \frac{2}{\alpha}) \right) \qquad (B.21)$$

where Γ denotes the Gamma function.

B.2.6 The Lognormal Distribution

A lognormal distribution relates closely to the normal distribution: one obtains a normal distribution if one takes the logarithm of samples of a lognormal distributed random variable. To define a lognormal distribution uniquely, we need two parameters, and usually one takes the parameters μ and σ that correspond to the mean and standard deviation of the normal distribution constructed as explained above.

A lognormal distributed random variable with parameters μ and σ has density

$$f(x) = \frac{1}{\sigma x \sqrt{(2\pi)}} e^{\frac{-(\ln(x)-\mu)^2}{2\sigma^2}}.$$

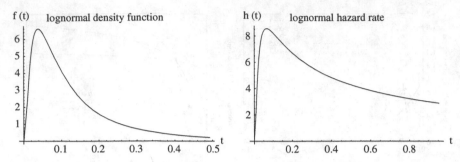

Fig. B.5 The probability density (*left*) and hazard rate (*right*) of a lognormal distribution ($\mu = -2.3, \sigma = 0.97$)

Appendix C
Estimating the Hazard Rate

C.1 Cumulative Hazard Rate

The cumulative hazard rate is estimated using the Nelson-Aalen estimator, which has especially good small sample performance. The Nelson-Aalen estimator is

$$
\hat{H}(t) = \begin{cases} 0 & \text{if } t \leq t_1 \\ \sum_{t_i \leq t} \frac{d_i}{Y_i} & \text{if } t_1 \leq t. \end{cases}
\tag{C.1}
$$

The estimated variance of the Nelson-Aalen estimator is

$$
\sigma_H^2(t) = \sum_{t_i \leq t} \frac{d_i}{Y_i^2}.
\tag{C.2}
$$

C.2 Epanechnikov Kernel

For the kernel $K(.)$ the Epanechnikov kernel is used

$$
K(x) = 0.75(1 - x^2) \qquad \text{for } -1 \leq x \leq 1
\tag{C.3}
$$

as it is shown in [80] to be often more accurate than other kernel functions. When $t-b < 0$ or $t+b > t_D$ the symmetric kernel must be transformed into an asymmetric one, which is at the lower bound with $q = t/b$

$$
K_q(x) = K(x)(\alpha + \beta x), \qquad \text{for } -1 \leq x \leq q,
\tag{C.4}
$$

where

$$
\alpha = \frac{64(2 - 4q + 6q^2 - 3q^3)}{(1+q)^4(19 - 18q + 3q^2)}
\tag{C.5}
$$

$$
\beta = \frac{240(1 - q)^2}{(1+q)^4(19 - 18q + 3q^2)}
\tag{C.6}
$$

For time-points in the right-hand tail $q = (t_D - 1)/b$ the kernel function is $K_q(-x)$.

C.3 Bandwidth Estimation

The *mean integrated squared error* (MISE) of the estimated hazard rate $\hat{h}$ over the range $\tau_{\min}$ to $\tau_{\max}$ is defined by

$$
\begin{aligned}
MISE(b) &= E\left(\int_{\tau_{\min}}^{\tau_{\max}} [\hat{h}(u) - h(u)]^2 \, du\right) \\
&= E\left(\int_{\tau_{\min}}^{\tau_{\max}} \hat{h}^2(u) \, du\right) - 2E\left(\int_{\tau_{\min}}^{\tau_{\max}} \hat{h}(u)h(u) \, du\right) \\
&+ E\left(\int_{\tau_{\min}}^{\tau_{\max}} h^2(u) \, du\right).
\end{aligned}
\tag{C.7}
$$

This function depends on the bandwidth b used in the Epanechnikov kernel. The last term does not contain b and can be ignored when finding the best value of b. The first term is estimated by $\int_{\tau_{\min}}^{\tau_{\max}} \hat{h}^2(u) \, du$. We evaluate $\hat{h}(u)$ at a not necessarily equidistant grid of points $\tau_{\min} = u_1 < u_2 < \ldots < u_M = \tau_{\max}$ and apply the trapezoid rule. The second term we approximate by a cross-validation estimate suggested by Ramlau-Hansen where we sum over the event times between $\tau_{\min}$ and $\tau_{\max}$.

Appendix D
The Laplace and the Laplace-Stieltjes Transform

This summary is taken from [147]. The Laplace transform (LT) and Laplace-Stieltjes transform (LST) apply to functions $F(t)$ satisfying $F(t) = 0$ for $t > 0$, $F(0)$ is analytical and

$$\int_{0^-}^{\infty} |F(t)| \, e^{-ct} \, dt < \infty \qquad \forall c > 0 \Rightarrow \lim_{t \to \infty} F(t) \, e^{-ct} = 0 \qquad \forall c. \qquad \text{(D.1)}$$

The Laplace transform $F^*(t)$ of $F(t)$ is defined as

$$F^*(s) = \int_{0^-}^{\infty} F(t) \, e^{-st} \, dt \qquad \text{(D.2)}$$

and the Laplace-Stieltjes transform is defined as

$$F^{\sim}(s) = \int_{0^-}^{\infty} e^{-st} \, dF(t). \qquad \text{(D.3)}$$

In [147] the relation of both transforms is shown to be

$$F^{\sim}(s) = s F^*(s). \qquad \text{(D.4)}$$

The Laplace-Stieltjes transform is of particular interest since it allows the manipulation of distributions when no analytical expression for the density exists.

The following table summaries the most important properties of the Laplace and the Laplace-Stieltjes transform.

Time domain	LT domain	LST domain
$F(t), t \geq 0$	$F^*(s) = \int\limits_{0^-}^{\infty} F(t)e^{-st}\, dt$	$F^{\sim}(s) = \int\limits_{0^-}^{\infty} e^{-st}\, dF(t)$
$aF(t) + bG(t)$	$aF^*(t) + bG^*(t)$	$aF^{\sim}(t) + bG^{\sim}(t)$
$F(\frac{t}{a}), a > 0$	$aF^*(as)$	$F^{\sim}(as)$
$F(t - a), a > 0$	$e^{-as}F^*(s)$	$e^{-as}F^{\sim}(s)$
$\int\limits_{0^-}^{t} F(\tau)G(t - \tau)\, d\tau$	$F^*(s)G^*(s)$	$\frac{1}{s}F^{\sim}(s)G^{\sim}(s)$
$\int\limits_{0^-}^{t} G(t - \tau)\, dF(\tau)$	$sF^*(s)G^*(s)$	$F^{\sim}(s)G^{\sim}(s)$
$\frac{dF(t)}{dt}$	$sF^*(s) - F(0)$	$s[F^{\sim}(s) - F(0)]$
$\int\limits_{0^-}^{\infty} F(t)\, dt$	$\lim\limits_{s\to 0} F^*(s)$	$\frac{1}{s}\lim\limits_{s\to 0} F^{\sim}(s)$
$\int\limits_{0^-}^{\infty} dF(t) = \lim\limits_{t\to\infty} F(t)$	$\lim\limits_{s\to 0} sF^*(s)$	$\lim\limits_{s\to 0} F^{\sim}(s)$
$\lim\limits_{t\to 0} F(t)$	$\lim\limits_{s\to\infty} sF^*(s)$	$\lim\limits_{s\to\infty} F^{\sim}(s)$

References

1. M. Abramowitz, I.A. Stegun (eds.), *Handbook of Mathematical Functions with Formulas, Graphs, and Mathematical Tables*, 9th edn. (Dover, New York, 1972)
2. E.N. Adams, Optimizing preventive service of software products. *IBM journal on Research and Development* **28**(1), 2–14 (1984)
3. M. Ajmone Marsan, G. Balbo, G. Conte, S. Donatelli, *Modelling with Generalized Stochastic Petri Nets*. Series in Parallel Computing (Wiley, New York, 1995)
4. M. Allman, V. Paxson, *On Estimating End-to-End Network Path Properties*. In *SIG-COMM'99: Proceedings of the Conference on Applications, Technologies, Architectures, and Protocols for Computer Communication*, New York, September 1999 (ACM Press, New York, 1999), pp. 263–274
5. H. Alt, L. Guibas, K. Mehlhorn, R. Karp, A. Wigderson, A method for obtaining randomized algorithms with small tail probabilities. *Algorithmica* **16**(4/5), 543–547 (1996)
6. J.R. Artalejo, A. Gómez-Corral, *Retrial Queueing Systems. A Computational Approach* (Springer, Berlin, 2008)
7. A. Avizienis, N-version approach to fault-tolerant software. *IEEE Trans. Software Eng.* **11**(12), 1491–1501 (1985)
8. A. Avritzer, E.J. Weyuker, Monitoring smoothly degrading systems for increased dependability. *Empirical Software Eng.* **2**(1), 59–77 (1997)
9. F. Baccelli, Analysis of a service facility with periodic checkpointing. *Acta Informatica* **15**, 67–81 (1981)
10. BEA Systems, IBM, Microsoft Corporation Inc, TIBCO Software Inc. Web Services Reliable Messaging Protocol (WS-Reliable Messaging), February 2005
11. M. Bertier, O. Marin, P. Sens, *Implementation and Performance Evaluation of an Adaptable Failure Detector*. In *DSN'02: Proceedings of the International Conference on Dependable Systems and Networks*, Bethesda, MD (IEEE Computer Society, Washington, DC, 2002), pp. 354–363
12. A. Bobbio, S. Garg, M. Gribaudo, A. Horvath, M. Sereno, M. Telek, *Modeling Software Systems with Rejuvenation, Restoration and Checkpointing Through Fluid Stochastic Petri Nets*. In *PNPM'99: Proceedings of the International Conference on Petri Nets and Performance Models, Zaragoza*, Spain, September 1999 (IEEE Computer Society, Los Alamitos, CA, 1999), pp. 82–91
13. A. Bobbio, A. Puliafito, M. Telek, K. Trivedi, Recent developments in non-Markovian stochastic Petri nets. *J. Syst. Circuits Comput.* **8**(1), 119–158 (1998)
14. A. Bobbio, M. Sereno, C. Anglano, Fine grained software degradation models for optimal rejuvenation policies. *Perform. Eval.* **46**(1), 45–62 (2001)
15. A. Bobbio, M. Telek, Chapter 7: The task completion time in degradable systems, in *Performability Modelling: Techniques and Tools*, ed. by B.R. Haverkort, R. Marie, G. Rubino, K.S. Trivedi (Wiley, New York, 2001), pp. 139–161
16. A. Brock, An analysis of checkpointing. *ICL Tech. J.* **1**(3), 211–228 (1979)

17. G. Candea, A.B. Brown, A. Fox, D.A. Patterson, Recovery-oriented computing: Building multitier dependability. *IEEE Comput.* **37**(11), 60–67 (2004)

18. G. Candea, J. Cutler, A. Fox, Improving availability with recursive microreboots: A soft-state system case study. *Perform. Eval.* **56**(1–3), 213–248 (2004)

19. G. Candea, J. Cutler, A. Fox, R. Doshi, P. Garg, R. Gowda, *Reducing Recovery Time in a Small Recursively Restartable System*. In *DSV'02: Proceedings of the International Conference on Dependable Systems and Networks*, Bethesda, MD, June 2002 (IEEE Computer Society, Washington, DC, 2002)

20. G. Candea, S. Kawamoto, Y. Fujiki, G. Friedman, A. Fox, *Microreboot: A Technique for Cheap Recovery*. In *OSDI'04: Proceedings of the USENIX/ACM Symposium on Operating Systems Design & Implementation*, San Francisco, CA, December 2004 (USENIX Association, Berkely, CA)

21. V. Castelli, R.E. Harper, P. Heidelberger, S.W. Hunter, K.S. Trivedi, K. Vaidyanathan, W.P. Zeggert, Proactive management of software aging. *IBM J. Res. Dev.* **45**(2), 311–332 (2001)

22. P. Chalasani, S. Jha, O. Shehory, K. Sycara, *Query Restart Strategies for Web Agents*. In *Proceedings of the Agents98*, New York (AAAI Press, Menlo Park, CA, 1998)

23. K.M. Chandy, A survey of analytic models of rollback and recovery strategies. *Computer* **8**(5), 40–47 (1975)

24. K.M. Chandy, J.C. Browne, C.W. Dissly, W.R. Uhrig, Analytic models for rollback and recovery strategies in data base systems. *IEEE Trans. Software Eng.* **SE-1**(1), 100–110 (1975)

25. K.M. Chandy, C.V. Ramamoorthy, Rollback and recovery strategies for computer programs. *IEEE Trans. Comput.* **C-21**(6), 546–556 (1972)

26. N. Chapin, *Do We Know What Preventive Maintenance Is?* In *ICSM'00: Proceedings of the International Conference on Software Maintenance*, San Jose, CA (IEEE Computer Society, Washington, DC, 2000), pp. 15–17

27. W. Chen, S. Toueg, M.K. Aguilera, *On the Quality of Service of Failure Detectors*. In *DSN'00: Proceedings of the International Conference on Dependable Systems and Networks*, New York (IEEE Computer Society, Los Alamitos, CA, 2000), pp. 191–200

28. G. Ciardo, D. Nicol, K.S. Trivedi, Discrete-event simulation of fluid stochastic Petri nets. *IEEE Trans. Software Eng.* **25**(2), 207–217 (1999)

29. E. Cinlar, *Introduction to Stochastic Processes* (Prentice-Hall, Englewood Cliffs, NJ, 1975)

30. E.G. Coffman Jr., E.N. Gilbert, Optimal strategies for scheduling checkpoints and preventive maintenance. *IEEE Trans. Reliab.* **39**(1), 9–18 (1990)

31. T. Courtney, D. Daly, S. Derisavi, S. Gaonkar, M. Griffith, V. Lam, W.H. Sanders, *The Möbius Modeling Environment: Recent Developments*. In *QEST'04: Proceedings of the 1st International Conference on Quantitative Evaluation of Systems*, Enschede, The Netherlands, September 2004 (IEEE Computer Society, Washington, DC, 2004), pp. 328–329

32. J.T. Daly, *A Model for Predicting the Optimum Checkpoint Interval for Restart Dumps*. In *Proceedings of the International Conference on Computational Science*. Lecture Notes in Computer Science, vol. 2660 (Springer, Heidelberg, 2003), pp. 3–12

33. J.T. Daly, *A Strategy for Running Large Scale Applications Based on a Model that Optimizes the Checkpoint Interval for Restart Dumps*, ed. by P.M. Johnson. In *Proceeding of the 1st International Workshop on Software Engineering for High Performance Computing System Applications* (IEE Digest, Edinburgh, Scotland, May 2004), pp. 70–74

34. T. Dohi, K. Goseva-Popstojanova, K.S. Trivedi, *Analysis of Software Cost Models with Rejuvenation*. In *HASE'00: Proceedings of the 5th IEEE International Symposium on High Assurance Systems Engineering*, Albuquerque, NM (IEEE Computer Society, Los Alamitos, CA, 2000), pp. 25–34

35. T. Dohi, K. Goseva-Popstojanova, K.S. Trivedi, *Statistical Non-Parametric Algorithms to Estimate the Optimal Software Rejuvenation Schedule*. In *PRDC'00: Proceedings of the Pacific Rim International Symposium on Dependable Computing*, Los Angels, CA, December 2000 (IEEE Computer Society, Los Alamitos, CA, 2000), pp. 77–84

36. T. Dohi, N. Kaio, K.S. Trivedi, *Availability Models with Age-Dependent Checkpointing*. In *SRDS'02: Proceedings of the 21st IEEE Symposium on Reliable Distributed Systems*, Suita, Japan (IEEE Computer Society, Washington, DC, 2002), pp. 130–139

37. A. Duda, The effects of checkpointing on program execution time. *Inform. Process. Lett.* **6**(5), 221–229 (1983)
38. J.B. Dugan, R. van Buren, Reliability evaluation of fly-by-wire computer systems. *J. Syst. Software* **25**(1), 109–120 (1994)
39. E.N. Elnozahy, J.S. Plank, Checkpointing for peta-scale systems: A look into the future of practical rollback-recovery. *IEEE Trans. Dependable Sec. Comput.* **1**(2), 97–108 (2004)
40. M. Elnozahy, L. Alvisi, Y.-M. Wang, D.B. Johnson, A survey of rollback-recovery protocols in message-passing systems. *ACM Comput. Surv.* **34**(3), 375–408 (2002)
41. L. Falai, A. Bondavalli, *Experimental Evaluation of the QoS of Failure Detectors on Wide Area Network*. In *DSN'05: Proceedings of the International Conference on Dependable Systems and Networks*, Yokohama, Japan (IEEE Computer Society, Washington, DC, 2005), pp. 624–633
42. G. Falin, A survey of retrial queues. *Queueing Syst.* **7**(2), 127–167 (1990)
43. G. Falin, J.G.C. Templeton, *Retrial Queues*. Monographs on Statistics & Applied Probability, vol. 75 (Chapman & Hall, London, 1997)
44. A. Fox, D. Patterson, *When Does Fast Recovery Trump High Reliability*. In *EASY'02: Proceedings of the 2nd Workshop on Evaluating and Architecting System Dependability*, San Jose, CA, October 2002
45. S. Garg, Y. Huang, C. Kintala, K.S. Trivedi, *Time and Load Based Software Rejuvenation: Policy, Evaluation and Optimality*. In *Proceedings of the 1st Conference on Fault-Tolerant Systems*, IIT, Madras, December 1995
46. S. Garg, Y. Huang, C. Kintala, K.S. Trivedi, *Minimizing Completion Time of a Program by Checkpointing and Rejuvenation*. In *SIGMETRICS'96: Proceedings of the ACM SIGMETRICS International Conference on Measurement and Modeling of Computer Systems*, Philadelphia, PA (ACM Press, New York, 1996), pp. 252–261
47. S. Garg, A. Puliafito, M. Telek, K. Trivedi, *Analysis of Software Rejuvenation Using Markov Regenerative Stochastic Petri Net*. In *ISSRE'95: Proceedings of the 6th International Symposium on Software Reliability Engineering*, Toulouse, France, (ACM, New York, USA) October 1995, pp. 180–187
48. S. Garg, A. Puliafito, M. Telek, K. Trivedi, *On the Analysis of Software Rejuvenation Policies*. In *COMPASS'97: Proceedings of the 12th Annual Conference on Computer Assurance*, Gaithersberg, MD, June 1997
49. S. Garg, A. Puliafito, M. Telek, K. Trivedi, Analysis of preventive maintenance in transactions based software systems. *IEEE Trans. Comput.* **47**(1), 96–107 (1998)
50. S. Garg, A. van Moorsel, K. Vaidyanathan, K.S. Trivedi, *A Methodology for Detection and Estimation of Software Aging*. In *ISSRE'98: Proceedings of the International Symposium on Software Reliability Engineering*, Paderborn, Germany, November 1998 (IEEE Computer Society, Los Alamitos, CA, 1998), pp. 283–292
51. R. Geist, R. Reynolds, J. Westall, Selection of a checkpoint in a critical-task environment. *IEEE Trans. Reliab.* **37**(4), 395–400 (1988)
52. E. Gelenbe, On the optimum checkpoint interval. *J. ACM* **26**(2), 259–270 (1979)
53. E. Gelenbe, D. Derochette, Performance of rollback recovery systems under intermittent failures. *Commun. ACM* **21**(6), 493–499 (1978)
54. E. Gelenbe, M. Hernández, Optimum checkpoints with age dependent failures. Acta Informatica **27**, 519–531 (1990)
55. R. German, *Analysis of Stochastic Petri Nets with Non-Exponentially Distributed Firing Times*. PhD thesis, Technical University of Berlin, Berlin, 1994
56. R. German, *Performance of Communication Systems, Modeling with Non-Markovian Stochastic Petri Nets* (Wiley, New York, 2000)
57. I. Gertsbakh, *Reliability Theory, with Applications to Preventive Maintenance* (Springer, Berlin, 2000)
58. M. Gittens, Y. Kim, D. Godwin, *The Vital Few Versus the Trivial Many: Examining the Pareto Principle for Software*. In *COMPSAC'05: Proceedings of the 29th Annual International Computer Software and Applications Conference*, Edinburgh, Scotland (IEEE Computer Society, Washington, DC, 2005), pp. 179–185

59. C.P. Gomes, B. Selman, H. Kautz, *Boosting Combinatorial Search Through Randomization.* In *AAAI'98: Proceedings of the 15th National Conference on Artificial Intelligence*, Madison, WI (AAAI Press, Menlo Park, CA, 1998), pp. 431–437

60. A. Goyal, V.F. Nicola, A.N. Tantawi, K.S. Trivedi, Reliability of systems with limited repairs. IEEE Trans. Reliab. **R-36**(2), 202–207 (1987)

61. S.R. Graham, *Fault Tolerance in Networked Control Systems Through Real-Time Restarts.* Technical Report A256524, Storming Media, Pentagon Reports, July 2004

62. S.L. Graham, M. Snir, C.A. Patterson (eds.), *Getting Up to Speed: The Future of Supercomputing* (The National Academies Press, Washington, DC, 2004)

63. V. Grassi, L. Donatiello, S. Tucci, On the optimal checkpointing of critical tasks and transaction-oriented systems. Trans. Software Eng. **18**(1), 72–77 (1992)

64. J. Gray, *Why Do Computers Stop and What Can Be Done About It?* In *Proceedings of the 5th Symposium on Reliability in Distributed Software and Database Systems*, Los Angeles, CA, January 1986 (IEEE Computer Society, Los Alamitos, CA, 1986), pp. 3–12. Extended Version appeared as Technical Report 85.7, June 1985, PN87614, Tandem Computers

65. M. Gribaudo, A. Horvath, Fluid stochastic Petri nets augmented with flush-out arcs: A transient analysis technique. IEEE Trans. Software Eng. **28**(10), 944–955 (2002)

66. M. Gribaudo, M. Sereno, A. Horvath, A. Bobbio, Fluid stochastic Petri nets augmented with flush-out arcs: Modelling and analysis. Discrete Event Dyn. Syst.: Theory Appl. **11**, 97–117 (2001)

67. V.P. Guddeti, B.Y. Choueiry, *An Empirical Study of a New Restart Strategy for Randomized Backtrack Search.* In *CP'04: Proceedings of the Workshop on CSP Techniques with Immediate Application*, Toronto, Canada, 2004, pp. 66–82

68. M. Hamill, K. Goseva-Popstojanova, Common trends in software fault and failure data. IEEE Trans. Software Eng. **35**(4), 484–496 (2009)

69. B.R. Haverkort, *Performance of Computer Communication Systems: A Model-Based Approach* (Wiley, Chichester, 1998)

70. J. Hong, S. Kim, Y. Cho, Cost analysis of optimistic recovery model for forked checkpointing. *IEICE Trans. Inform. Syst.* **E-85A**(1), 1534–1541 (2002)

71. Y. Hong, D. Chen, L. Li, K.S. Trivedi, *Closed Loop Design for Software Rejuvenation.* In *SHAMAN'02: Proceedings of the Workshop on Self-Healing, Adaptive, and Self-Managed Systems (SHAMAN)*, New York, 2002

72. G. Horton, V.G. Kulkarni, D.M. Nicol, K.S. Trivedi, Fluid stochastic Petri nets: Theory, applications and solution. Eur. *J. Oper. Res.* **105**(1), 184–201 (1998)

73. Y. Huang, C. Kintala, N. Kolettis, N.D. Fulton, *Software Rejuvenation: Analysis, Module and Applications.* In *Proceedings of the 25th Symposium on Fault Tolerant Computing*, Pasadena, CA, June 1995 (IEEE Computer Society, Los Alamitos, CA, 1995), pp. 381–390

74. V. Jacobson, *Congestion Avoidance and Control.* In *Proceedings of the SIGCOMM'88 Symposium, Stanford, CA, August 1988; ACM Comput. Commun. Rev.* **18**(4), 314–329

75. N.K. Jaiswal. *Priority Queues* (Academic Press, New York, 1968)

76. S. Kalaiselvi, V. Rajaraman, A survey of checkpointing algorithms for parallel and distributed computers. *Sadhana* **25**(5), 489–510 (2000)

77. K. Kant, *A Global Checkpointing Model for Error Recovery.* In *AFIPS'83: Proceedings of the National Computer Conference, Anaheim*, CA, 16–19 May, 1983 (ACM Press, New York, 1983), pp. 81–89

78. S. Karlin, H.E. Taylor, *A First Course in Stochastic Processes* (Academic Press, New York, 1975)

79. P. Karn, C. Partridge, Improving round-trip time estimates in reliable transport protocols. *ACM Trans. Comput. Syst.* **9**(4), 364–373 (1991)

80. J.P. Klein, M.L. Moeschberger, *Survival Analysis, Techniques for Censored and Truncated Data* (Springer, New York, 1997)

81. R. Koo, S. Toueg, Checkpointing and rollback-recovery for distributed systems. IEEE Trans. *Software Eng.* **SE-13**(1), 23–31 (1987)

82. V.P. Koutras, A.N. Platis, *Semi-Markov Availability Modeling of a Redundant System with Partial and Full Rejuvenation Actions.* In *DEPCOS-RELCOMEX'08: Proceedings of the 3rd International Conference on Dependability of Computer Systems DepCoS-RELCOMEX*, Szklarska Poręba (IEEE Computer Society, Washington, DC, 2008), pp. 127–134

83. S. Kowshik, G. Baliga, S. Graham, L. Sha, *Co-Design Based Approach to Improve Robustness in Networked Control Systems.* In *DSN'05: Proceedings of the International Conference on Dependable Systems and Networks*, Kyoto, Japan (IEEE Computer Society, Los Alamitos, CA, 2005), pp. 454–463

84. C.M. Krishna, Y.-H. Lee, K.G. Shin, Optimization criteria for checkpoint placement. Commun. *ACM* **27**(10), 1008–1012 (1984)

85. C.M. Krishna, K.G. Shin, *Performance Measures for Multiprocessor Controllers.* In *Performance'83: Proceedings of the 9th International Symposium on Computer Performance Modelling, Measurement and Evaluation*, College Park, MD (North-Holland Publishing Co., Amsterdam, The Netherlands, 1983), pp. 229–250

86. B. Krishnamurthy, J. Rexford, *Web Protocols and Practice* (Addison-Wesley, Reading, MA, 2001)

87. V.G. Kulkarni, *Modeling and Analysis of Stochastic Systems* (Chapman & Hall, London, 1995)

88. V.G. Kulkarni, V.F. Nicola, R.M. Smith, K.S. Trivedi, *Numerical Evaluation of Performability and Job Completion Time in Repairable Fault-Tolerant Systems.* In *Proceedings of the 16th International Symposium on Fault-Tolerant Computing*, Vienna, Austria (IEEE Computer Society, Los Alamitos, CA, 1986), pp. 252–257

89. V.G. Kulkarni, V.F. Nicola, K.S. Trivedi, The completion time of a job on multimode systems. *Adv. Appl. Probab.* **19**, 923–954 (1987)

90. V.G. Kulkarni, V.F. Nicola, K.S. Trivedi, Effects of checkpointing and queueing on program performance. *Commun. Stochastic Models* **6**(4), 615–648 (1990)

91. A.M. Law, W.D. Kelton, *Simulation Modeling & Analysis*, 2nd edn. Industrial Engineering and Management Science (McGraw-Hill, New York, 1991)

92. P. L'Ecuyer, J. Malenfant, Computing optimal checkpointing strategies for rollback and recovery systems. *IEEE Trans. Comput.* **37**(4), 491–496 (1988)

93. C.H.C. Leung, Q.H. Choo, On the execution of large batch programs in unreliable computing systems. *IEEE Trans. Software Eng.* **10**(4), 444–450 (1984)

94. C.-C.J. Li, W.K. Fuchs, *CATCH-Compiler-Assisted Techniques for Checkpointing.* In *Proceedings of the 20th International Symposium on Fault-Tolerant Computing*, Newcastle Upon Tyne, June 1990 (IEEE Computer Society, Los Alamitos, CA, 1990), pp. 74–81

95. B.C. Ling, E. Kiciman, A. Fox, *Session State: Beyond Soft State.* In *NSDI'04: Proceedings of the 1st Symposium on Network Systems Design and Implementation (NSDI)*, San Francisco, CA, March 2004 (USENIX Association, Berkely, CA, 2004), pp. 295–308

96. Y. Liu, K.S. Trivedi, Y. Ma, J.J. Han, H. Levendel, *Modeling and Analysis of Software Rejuvenation in Cable Modem Termination Systems.* In *ISSRE'02: Proceedings of the 13th International Symposium on Software Reliability Engineering*, Annapolis, MD, November 2002 (IEEE Computer Society, Los Alamitos, CA, 2002), pp. 159–170

97. N. Looker, J. Xiu, *Assessing the Dependability of Soap-Rpc-Based Web Services by Fault Injection.* In *Proceedings of the 9th IEEE International Workshop on Object-oriented Real-time Dependable Systems*, Anacapri, Italy (IEEE Computer Society, Washington, DC, 2003), pp. 163–170

98. D.E. Lowell, S. Chandra, P.M. Chen, *Exploring Failure Transparency and the Limits of Generic Recovery.* In *OSDI'00: Proceedings of the 4th USENIX Symposium on Operating Systems Design and Implementation*, San Diego, CA, October 2000 (USENIX Association, Berkely, CA, 2000), pp. 289–303

99. M. Luby, A. Sinclair, D. Zuckerman, *Optimal Speedup of Las Vegas Algorithms.* In *ISTCS'93: Proceedings of the Israel Symposium on Theory of Computing Systems*, Natanya, Israel (IEEE Computer Society, Los Alamitos, CA, 1993), pp. 128–133

100. M. Lyu, *A Design Paradigm for Multi-Version Software Mathematics Subject Classification*. PhD thesis, University of California, Los Angeles, CA, 1988

101. M. Lyu, A. Avizienis, *Assuring Design Diversity in N-Version Software: A Design Paradigm for N-Version Programming*, ed. by J.F. Meyer, R.D. Schlichting. In *Proceedings of the Dependable Computing and Fault-Tolerant Systems* (Springer, Vienna, Austria, 1992), pp. 197–218

102. M.J. Magazine, Optimality of intuitive checkpointing policies. *Inform. Process. Lett.* **17**(2), 63–66 (1983)

103. R. Matias Jr., P.J.F. Filho, *An Experimental Study on Software Aging and Rejuvenation in Web Servers*. In *COMPSAC'06: Proceedings of the 30th Annual International Computer Software and Applications Conference*, Chicago, IL (IEEE Computer Society, Washington, DC, 2006), pp. 189–196

104. S.M. Maurer, B.A. Huberman, Restart strategies and internet congestion. J. Econ. Dyn. Control **25**, 641–654 (2001)

105. D.F. McAllister, M.A. Vouk, Fault-tolerant software reliability engineering, in *Handbook of Software Reliability Engineering*, ed. by M. Lyu (McGraw-Hill, New York, 1996), pp. 567–614

106. J.J. McCall, Maintenance policies for stochastically failing equipment: A survey. Manage. Sci. **11**, 493–521 (1965)

107. D.S. Mitrinovic, *Analytic Inequalities* (Springer, New York, 1970)

108. R. Motwani, P. Raghavan, *Randomized Algorithms* (Cambridge University Press, New York, 1995)

109. H. Naik, R. Gupta, P. Beckman, *Analyzing Checkpointing Trends for Applications on the IBM Blue Gene/p System*. Technical Report Preprint ANL/MCS-P1590-0309, Argonne National Laboratory, Argonne, IL, March 2009

110. H.G. Naik, R. Gupta, P. Beckman, *Analyzing Checkpointing Trends for Applications on Peta-Scale Systems*. In *P2S2'09: Proceedings of the 2nd International Workshop on Parallel Programming Models and Systems Software (P2S2) for High-End Computing* (IEEE Computer Society, Vienna, Austria, 2009)

111. M.F. Neuts, *Matrix-Geometric Solutions in Stochastic Models: An Algorithmic Approach* (The Johns Hopkins University Press, Baltimore, MD, 1981)

112. V.F. Nicola, P. Shahabuddin, M. Nakayama, Techniques for the fast simulation of models of highly dependable systems. *IEEE Trans. Reliab.* **50**, 246–264 (2001)

113. V.F. Nicola, Chapter 7: Checkpointing and the modeling of program execution time, in *Software Fault Tolerance*, ed. by M.R. Lyu. *Trends in Software*, vol. 3 (Wiley, Chichester, 1995), pp. 167–188

114. V.F. Nicola, A. Bobbio, K.S. Trivedi, A unified performance reliability analysis of a system with a cumulative down time constraint. *Microelectron. Reliab.* **32**(1/2), 49–65 (1992)

115. A.J. Oliner, L. Rudolph, R. Sahoo, *Cooperative Checkpointing Theory*. In *Proceedings of the Parallel and Distributed Processing Symposium*, Rhodes Island, Greece, January 2006 (IEEE Computer Society, Los Alamitos, CA/ACM Press, 2006)

116. A.J. Oliner, R. Sahoo, *Evaluating Cooperative Checkpointing for Supercomputing Systems*. In *SMTPS'06: Proceedings of the 2nd Workshop on System Management Tools for Large-Scale Parallel Systems at IPDPS*, Greece, January 2006 (IEEE Computer Society, Los Alamitos, CA/ACM Press, 2006)

117. T.J. Ostrand, E.J. Weyuker, The distribution of faults in a large industrial software system. *ACM SIGSOFT Software Eng. Notes* **27**(4), 55–64 (2002)

118. V. Paxson, *End-to-End Internet Packet Dynamics*. In *Proceedings of the ACM SIGCOMM'97 Conference on Applications, Technologies, Architectures, and Protocols for Computer Communication*, Cannes, France, September 1997 (ACM Press, New York, 1997); Comput. Commun. Rev. **27**(4), 139–154

119. V. Paxson, *On Calibrating Measurements of Packet Transit Times*. In *SIGMETRICS'98/PERFORMANCE '98: Proceedings of the ACM SIGMETRICS Joint International Conference on Measurement and Modeling of Computer Systems*, Madison, WI (ACM Press, New York, 1998), pp. 11–21

120. V. Paxson, M. Allmann, Computing TCP's retransmission timer (RFC 2988, IETF, Nov 2000), http://www.rfc-editor.org/rfc/rfc2988.txt

121. A. Pfening, S. Garg, M. Telek, A. Puliafito, K.S. Trivedi, Optimal software rejuvenation for tolerating soft failures. *Perform. Eval.* **27, 28**, 491–506 (1996)

122. W.P. Pierskalla, J.A. Voelker, A survey of maintenance models: The control and surveillance of deteriorating systems. *Naval Res. Logist. Quart.* **23**, 353–388 (1976)

123. M. Poniatowski, *UNIX User's Handbook* (Prentice-Hall, Upper Saddle River, NJ, 2001)

124. D.K. Pradhan, N.H. Vaidya, Roll-forward and rollback recovery: Performance-reliability trade-off. *IEEE Trans. Comput.* **46**(3), 372–378 (1997)

125. F. Quaglia, A cost model for selecting checkpoint positions in time warp parallel simulation. *IEEE Trans. Parall. Distr. Syst.* **12**(4), 346–362 (2001)

126. B. Randell, System structure for software fault tolerance. *IEEE Trans. Software Eng.* **SE-1**(2), 221–232 (1975)

127. B. Randell, P.A. Lee, P.C. Treleaven, Reliability issues in computing system design. *ACM Comput. Surv.* **10**(2), 123–165 (1978)

128. P. Reinecke, A. van Moorsel, K. Wolter, *A Measurement Study of the Interplay Between Application Level Restart and Transport Protocol*. In *ISAS'04: Proceedings of the International Service Availability Symposium*, Munich, Germany, May 2004. Lecture Notes in Computer Science, vol. 3335 (Springer, Heidelberg, 2004)

129. P. Reinecke, A.P.A. van Moorsel, K. Wolter, *Experimental Analysis of the Correlation of HTTP GET Invocations*, ed. by M. Telek, A. Horvath. In *Proceedings of the 3rd European Performance Engineering Workshop*, Budapest, Hungary. Lecture Notes in Computer Science, vol. 4054 (Springer, Heidelberg, 2006)

130. P. Reinecke, A.P.A. van Moorsel, K. Wolter, *The Fast and the Fair: A Fault-Injection-Driven Comparison of Restart Oracles for Reliable Web Services*. In *QEST'06: Proceedings of the 3rd International Conference on the Quantitative Evaluation of Systems*, Riverside, CA, September 2006 (IEEE Computer Society, Los Alamitos, CA, 2006)

131. K. Rinsaka, T. Dohi, *Non-parametric Predictive Inference of Preventive Rejuvenation Schedule in Operational Software Systems*. In *ISSRE'07: Proceedings of the 18th IEEE International Symposium on Software Reliability*, Trollhättan, Sweden (IEEE Computer Society, Washington, DC, 2007), pp. 247–256

132. Y. Ruan, E. Horvitz, H. Kautz, *Restart Policies with Dependence Among Runs: A Dynamic Programming Approach*. In *Proceedings of the 8th International Conference on Principles and Practice of Constraint Programming*, Ithaca, NY, September 2002. Lecture Notes in Computer Science, vol. 2470 (Springer, Heidelberg, 2002), pp. 573–586

133. F. Salfner, G.A. Hoffmann, M. Malek, in *Prediction-Based Software Availability Enhancement*, ed. by O. Babaoglu, M. Jelasity, A. Montresor, C. Fetzer, S. Leonardi, A. van Moorsel, M. van Steen. Self-Star Properties in Complex Information Systems. Lecture Notes in Computer Science, vol. 3460 (Springer, Heidelberg, 2005)

134. W.H. Sanders, J.F. Meyer, in *Stochastic Activity Networks: Formal Definitions and Concepts*, ed. by E. Brinksma, H. Hermanns, J.P. Katoen. Lectures on Formal Methods and Performance Analysis, First EEF/Euro Summer School on Trends in Computer Science, Berg en Dal, The Netherlands, July 2000. Revised Lectures Series: Lecture Notes in Computer Science, vol. 2090 (Springer, Heidelberg, 2000, 2001), pp. 315–343

135. S. Sankaran, J.M. Squyres, B. Barrett, A. Lumsdaine, J. Duell, P. Hargrove, E. Roman, The LAM/MPI checkpoint/restart framework: System-initiated checkpointing. *Int. J. High Perform. Comput. Appl.* **19**(4), 479–493 (2005)

136. M. Schroeder, L. Buro, *Does the Restart Method Work? Preliminary Results on Efficiency Improvements for Interactions of Web-Agents*, ed. by T. Wagner, O. Rana. In *Proceedings of the Workshop on Infrastructure for Agents, MAS, and Scalable MAS at the Conference Autonomous Agents*, Montreal, Canada (Springer, Heidelberg, 2001)

137. K.G. Shin, T.H. Lin, Y.H. Lee, Optimal checkpointing of real-time tasks. *IEEE Trans. Comput.* **C-36**(11), 1328–1341 (1987)

138. D.P. Siewiorek, R.S. Swarz, *Reliable Computer Systems: Design and Evaluation*, 2nd edn. (Digital Press, Newton, MA, 1982)

139. L.M. Silva, J. Alonso, J. Torres, Using virtualization to improve software rejuvenation. *IEEE Trans. Comput.* **58**(11), 1525–1538 (2009)

140. L.M. Silva, J.G. Silva, *System-Level Versus User-Defined Checkpointing*. In *SRDS'98: Proceedings of the 17th IEEE Symposium on Reliable Distributed Systems*, West Lafayette, IN (IEEE Computer Society, Washington, DC, 1998), pp. 68–74

141. Sprint. *http://www.sprintworldwide.com/english/solutions/sla/*. Last accessed Jan 2006

142. V. Sundaram, S.H. Chaudhuri, S. Garg, C. Kintala, S. Bagchi, *Improving Dependability Using Shared Supplementary Memory and Opportunistic Micro Rejuvenation in Multi-Tasking Embedded Systems*. In *PRDC'07: Proceedings of the 13th Pacific Rim International Symposium on Dependable Computing*, Melbourne, VIC, Australia (IEEE Computer Society, Washington, DC, 2007), pp. 240–247

143. A.T. Tai, J.F. Meyer, A. Avizienis, *Software Performability: From Concepts to Applications*. Springer International Series in Engineering and Computer Science, vol. 347 (Springer, Berlin, 1996)

144. A.T. Tai, K.S. Tso, *A Performability-Oriented Software Rejuvenation Framework for Distributed Applications*. In *DSN'05: Proceedings of the International Conference on Dependable Systems and Networks*, Yokohama, Japan (IEEE Computer Society, Washington, DC, 2005), pp. 570–579

145. A.S. Tanenbaum, *Computer Networks* (Prentice-Hall, Upper Saddle River, NJ, 1996)

146. A.N. Tantawi, M. Ruschitzka, Performance analysis of checkpointing strategies. *ACM Trans. Comput. Syst.* **2**(2), 123–144 (1984)

147. M. Telek, *Some Advanced Reliability Modelling Techniques*. PhD thesis, Technical University of Budapest, Hungary, 1994

148. S. Thanawastien, R.S. Pamula, Y.L. Varol, *Evaluation of Global Checkpoint Rollback Strategies for Error Recovery in Concurrent Processing Systems*. In *Proceedings of the 16th International Symposium on Fault-Tolerant Computing*, New York (IEEE Computer Society, Washington, DC, 1986), pp. 246–251

149. The Apache Software Foundation: Apache Axis, http://ws.apache.org/axis/

150. The Apache Software Foundation: Apache Sandesha, http://ws.apache.org/sandesha/

151. T. Thein, J.S. Park, Availability analysis of application servers using software rejuvenation and virtualization. *J. Comput. Sci. Technol.* **24**(2), 339–346 (2009)

152. S. Toueg, Ö. Babaoglu, On the optimum checkpoint selection problem. *SIAM J. Comput.* **13**(3), 630–649 (1984)

153. K. Trivedi, K. Vaidyanathan, K. Goseva-Popstojanova, *Modeling and Analysis of Software Aging and Rejuvenation*. In *Proceedings of the 33rd Annual Simulation Symposium*, Washington, DC (IEEE Computer Society, Los Alamitos, CA, 2000), pp. 270–279

154. K.S. Trivedi, *Probability and Statistics with Reliability, Queuing, and Computer Science Applications* (Wiley, New York, 2001)

155. K.S. Trivedi, V.G. Kulkarni, *FSPNs: Fluid Stochastic Petri Nets*. In *Proceedings of the 14th International Conference on the Application and Theory of Petri Nets*, Chicago, IL. Lecture Notes in Computer Science, vol. 691 (Springer, Heidelberg, 1993), pp. 24–31

156. N.H. Vaidya, Impact of checkpoint latency on overhead ratio of a checkpointing scheme. *IEEE Trans. Comput.* **46**(8), 942–947 (1997)

157. K. Vaidyanathan, R.E. Harper, S.W. Hunter, K.S. Trivedi, Analysis and implementation of software rejuvenation in cluster systems. *SIGMETRICS Perform. Eval. Rev.* **29**(1), 62–71 (2001)

158. K. Vaidyanathan, D. Selvamuthu, K.S. Trivedi, *Analysis of Inspection-Based Preventive Maintenance in Operational Software Systems*. In *SRDS'02: Proceedings of the 21st Symposium on Reliable Distributed Systems*, Osaka, Japan (IEEE Computer Society, Washington, DC, 2002), pp. 286–295

159. K. Vaidyanathan, K.S. Trivedi, *A Measurement-Based Model for Estimation of Resource Exhaustion in Operational Software Systems*. In *ISSRE'99: Proceedings of the 10th International Symposium on Software Reliability Engineering*, Boca Raton, FL, November 1999 (IEEE Computer Society, Washington, DC, 1999), pp. 84–93

160. C. Valedez-Flores, R.M. Feldman, A survey of preventive maintenance models for stochastically deteriorating single-unit systems. *Naval Res. Logist.* **36**, 419–446 (1989)
161. A.P.A. van Moorsel, K. Wolter, *Optimization of Failure Detection Retry Times*. In *DSN'03: Proceedings of the International Conference on Dependable Systems and Networks*, San Francisco, CA (IEEE Computer Society, Washington, DC, 2003). (Fast Abstract)
162. A.P.A. van Moorsel, K. Wolter, *Optimization of Failure Detection Retry Times*. In *Proceedings of the Performability Workshop*, Monticello, IL, September 2003
163. A.P.A. van Moorsel, K. Wolter, *Analysis and Algorithms for Restart*. In *QEST'04: Proceedings of the 1st International Conference on the Quantitative Evaluation of Systems*, Twente, The Netherlands, September 2004 (IEEE Computer Society, Los Alamitos, CA, 2004), pp. 195–204. (Best paper award)
164. A.P.A. van Moorsel, K. Wolter, *Making Deadlines Through Restart*. In *MMB'04: Proceedings of the 12th GI/ITG Conference on Measuring, Modelling and Evaluation of Computer and Communication Systems*, Dresden, Germany, September 2004 (VDE, Berlin, 2004), pp. 155–160
165. A.P.A. van Moorsel, K. Wolter, *Optimal Restart Times for Moments of Completion Time*. In *UKPEW'04: Proceedings of the UK Performance Engineering Workshop*, Bradford, July 2004. (Selected for a special issue of IEE Proc. Software J.)
166. A.P.A. van Moorsel, K. Wolter, *A Short Investigation into an Underexplored Model for Retries*. In PMCCS-7: Proceedings of the 7th International Workshop on Performability of Computer and Communication Systems, University of Torino, Torino, Italy, September 2005
167. A.P.A. van Moorsel, K. Wolter, Analysis of restart mechanisms in software systems. *IEEE Trans. Software Eng.* **32**(8), 547–558 (2006)
168. A.P.A. van Moorsel, K. Wolter, Optimal restart times for moments of completion time. *IEE Proc. Software* **151**(5), 219–223 (2004)
169. J. Villén-Altamirano, Rare event restart simulation of two-stage networks. *Eur. J. Oper. Res.* **179**(1), 148–159 (2007)
170. J.L. von Neumann, Probabilistic logics and the synthesis of reliable organisms from unreliable components. *Automata Stud.* **34**, 43–98 (1956)
171. T. Walsh, *Search in a Small World*. In *IJCAI: Proceedings of the International Joint Conference on AI (IJCAI)*, IJCAII and the Scandinavian AI Societies, San Franscisco, 1999, pp. 1172–1177
172. L. Wang, K. Pattabiraman, Z. Kalbarczyk, R.K. Iyer, L. Votta, C. Vick, A. Wood, *Modeling Coordinated Checkpointing for Large-Scale Supercomputers*. In *DSN'05: Proceedings of the Dependable Systems and Networks*, Yokohama, Japan (IEEE Computer Society, Washington, DC, 2005), pp. 812–821
173. Y.-M. Wang, Y. Huang, K.-P. Vo, P.-Y. Chung, C.M.R. Kintala, *Checkpointing and Its applications*. In *FTCS-25: Proceedings of the 25th International Symposium on Fault-Tolerant Computing*, Pasadena, CA, June 1995 (IEEE Computer Society, Los Alamitos, CA, 1995), pp. 22–31
174. H. White, L.S. Christie, Queuing with preemptive priorities or with breakdown. *Oper. Res.* **6**(1), 79–95 (1958)
175. K. Wolter, *Second Order Fluid Stochastic Petri Nets: An Extension of GSPNs for Approximate and Continuous Modeling*. In *WCSS'97: Proceedings of the 1st World Congress on Systems Simulation*, Society of Computer Simulation, Singapore, 1–3 September 1997, pp. 328–332
176. K. Wolter, *Jump Transitions in Second Order FSPNs*. In *MASCOTS'99: Proceedings of the 7th International Symposium on Modelling, Analysis and Simulation of Computer and Telecommunication Systems*, College Park, MD, October 1999 (IEEE Computer Society, Washington, DC, 1999), pp. 156–163
177. K. Wolter, in *Self-Management of Systems Through Automatic Restart. Self-Star Properties in Complex Information Systems*. Lecture Notes in Computer Science, vol. 3460 (Springer, Heidelberg, 2005), pp. 189–203

178. K. Wolter, A.P.A. van Moorsel, in *Self-Management of Systems Through Automatic Restart. SELF-STAR: International Workshop on Self-Star Properties in Complex Information Systems*, University of Urbino, Bertinoro, Italy, June 2004

179. K. Wolter, A. Zisowsky, On Markov reward modelling with FSPNs. *Perform. Eval.* **44**, 165–186 (2001)

180. J. Xu, B. Randell, *Roll-Forward Error Recovery in Embedded Real-Time Systems*. In *ICPADS'96: Proceedings of the International Conference on Parallel and Distributed Systems*, Tokyo, Japan (IEEE Computer Society, Washington, DC, 1996), pp. 414–421

181. J.W. Young, A first order approximation to the optimum checkpoint interval. *Commun. ACM* **17**(9), 530–531 (1974)

182. A. Zimmermann, J. Freiheit, R. German, G. Hommel, *Petri Net Modelling and Performability Evaluation with TimeNET 3.0*. In *TOOLS'00: Proceedings of the 11th International Conference on Computer Performance Evaluation: Modelling Techniques and Tools* (Springer, London, 2000), pp. 188–202

183. A. Ziv, J. Bruck, An on-line algorithm for checkpoint placement. *IEEE Trans. Comput.* **46**(9), 976–985 (1997)

Index

Glossary

CDF	*cumulative distribution function* also named probability distribution function. See also PDF compression factor ρ factor of proportionality in reprocessing the audit trail equals utilisation of the queue. *Page 209*
CTMC	continuous-time Markov chain. *Page 147*
DSPN	deterministic and stochastic Petri net. *Page 146*
DTMC	discrete-time Markov chain. *Page 139*
ER	emergency repair. *Page 125*
ETT	Effective Transmission Time. *Page 82*
FSPN	fluid stochastic Petri net. *Page 149*
γ	failure rate
GSPN	generalised stochastic Petri net. *Page 146*
iid	independent and identically distributed. *Page 13*
kurtosis	β2, normalized form of the fourth central moment which describes the degree of peakedness of a distribution. *Page 53*
kurtosis excess	γ2, scaled kurtosis, such that the normal distribution has kurtosis excess zero. *Page 53*
λ	transaction arrival rate and scale parameter of the Weibull distribution.
LST	Laplace-Stieltjes transform
MRGP	Markov-regenerative process. *Page 138*
MRSPN	Markov-regenerative stochastic Petri net. *Page 149*
MTBF	mean time between failures. *Page 119*
MTTF	mean time to failure. *Page 129*
MTTR	mean time to repair. *Page 129*

μ	task processing rate
ν	repair rate
PDF	*Probability Distribution Function* also named cumulative distribution function. See also CDF pdf Probability density function
PM	preventive maintenance. *Page 125*
Preemptive repeat different failure (prd)	Failure mode where the performed work is lost the task is reprocessed in stochastic different sense. *Page 7*
Preemptive repeat identical failure (pri)	Failure mode where the performed work is lost and the task is reprocessed in stochastic identical sense. *Page 7*
Preemptive resume failure (prs)	Failure mode where the performed work is saved and only the remaining work must be processed. *Page 7*
Program level checkpointing	checkpoint, rollback and recovery of a long-running task. *Page 7*
R	recovery after failure, or rejuvenation, deterministic or random variable. *Page 139*
Reboot	Possible update or reinstallation of application, restart and retry of tasks. *Page 8*
Restart	Reload of an application and reprocessing of tasks. *Page 7*
Retry	Repetition of stochastically identical request or task. *Page 7*
RTO	retransmission timeout. *Page 40*
RTT	round-trip time. *Page 40*
SAN	stochastic activity net. *Page 151*
skewness	third central moment expressing the steepness of the slope of a distribution. *Page 53*
SLA	service-level agreement. *Page 40*
SMP	semi-Markov process. *Page 152*
SRN	stochastic reward net. *Page 149*

System level checkpointing   saving the system state in a checkpoint and the
                             request list in an audit trail, rollback and recovery of
                             the system state, replay of the audit trail. *Page 8*

TCP   transmission control protocol. *Page 40*

URC   Unnecessary Ressource Consumption. *Page 83*